THE TRANSPORT ENGINEER'S HANDBOOK 1984

You need it, we've got it.

A vehicle for every business. You'll find Bedfords working in just about every business you can think of. That's because we offer a vast range of vehicles from light vans to heavy-duty trucks, to keep the wheels of British industry turning.

A dealer round every corner. You'll find a Bedford dealer is never far away. He'll help you choose the vehicles that best suit your business needs, advise you on our special leasing, insurance and warranty schemes and he'll maintain your Bedfords in peak condition throughout their long working life.

BEDFORD MEANS BUSINESS

BEDFORD COMMERCIAL VEHICLES. P.O. BOX 3. LUTON. BEDS. LU2 0SY.

THE TRANSPORT ENGINEER'S HANDBOOK 1984

Consultant Editor:
Graham Montgomerie

This (second) edition first published 1984
by Kogan Page Ltd, 120 Pentonville Road,
London N1 9JN

Copyright © Kogan Page Ltd 1984
All rights reserved

British Library Cataloguing in Publication Data
The transport engineer's handbook

1. Motor vehicles – Handbooks, manuals, etc. – Periodicals
629.2'005   TL147
ISBN 0-85038-793-0

Printed in Great Britain by Anchor Press
and bound by Wm Brendon and Son, both
of Tiptree, Essex

GIVE YOUR ENGINES THE OIL THAT'S RIGHT FOR THE WORK THEY DO

FROM CASTROL

Choosing the correct oil for an engine is vital particularly for the fleet operator. Oil has a major influence on performance of vehicles and length of their working life. The correct choice of oil will minimise operating costs.

That is why Castrol has recognised the different needs of each type of operation and provides oils formulated to meet your specific requirements.

Check the table to see which oil you need for your vehicle category. It will help you to keep down maintenance costs and ensure a longer, more profitable working life for your vehicles.

For further information and technical advice, please contact Geoff Wheeler, Industrial Lubricants Division, Burmah-Castrol Ltd., Burmah House, Pipers Way, Swindon, Wiltshire. SN3 1RE. Tel: Swindon (0793) 30151. Ext. 2176.

Vehicle Category	Recommended Grade
General Haulage vehicles	Deusol RX Super 15W/40
Light Commercial Vehicles	Deusol CRX 10W/40
Passenger Service Vehicles	Mershol MG 15W/40
Off-Highway Vehicles & Plant	Deusol Multiplant

Contents

Introduction	9
Part 1	**11**
1. Recent Technical Developments	13
2. Diesel Waxing and its Prevention	37
3. The Design and Construction of Vehicles for the Carriage of Dangerous Substances	51
4. Vehicle Noise Legislation	65

Rubber and Air Suspension Systems for Rigid Vehicles and Trailer Running Gears

Singles   10-13 tonnes
Tandems   18-40 tonnes
Triaxles and Third Axles to suit new legislation
- Long life, low maintenance
- Premium quality, bespoke engineering
- 24 month, 250,000km warranty
- High articulation, fail safe design
- Low deck height ability
- 'Specials' customer engineering capacity

NORDE Suspensions Ltd.

SUSPENSIONS

Sywell Airport, Northampton NN6 0BU England
Telephone: Northampton (0604) 493161
Telex: 311400

An Aid To Economic Driving

WHATEVER type of operation, fuel economy is foremost in the mind of vehicle operators. The answer is to fit a Lucas Kienzle Top Speed Limiter to achieve:

- Fuel savings by reduction in MAXIMUM speed.
- Full benefit from power saving devices, e.g. air deflectors, radiator shutters etc.
- Reduction in wear, tear and therefore maintenance.
- Improved safety by reducing maximum speed, resulting in lower insurance costs.

Lucas Kienzle's Top Speed Limiter features the following benefits:

- No air supply required.
- Electronic control with electric actuation for excellent driveability.
- Tight speed regulation so that changes are so gradual they are virtually undetectable.
- Once fitted the system is totally maintenance-free

Lucas Kienzle Instruments Limited
36 Gravelly Industrial Park, Birmingham B24 8TA. Tel: 021-328 5533 Telex: 335563

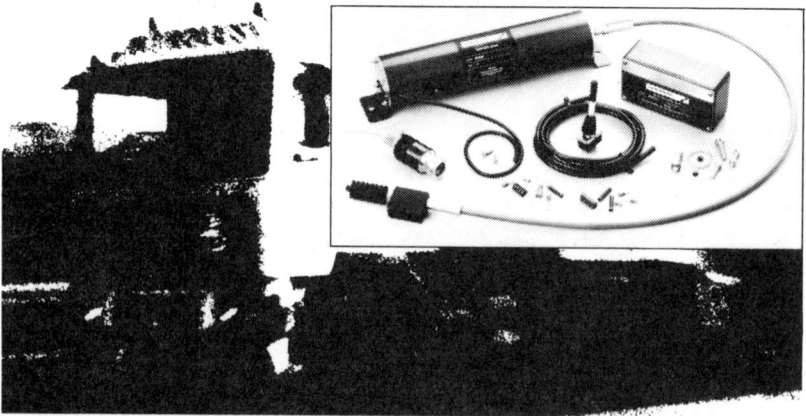

5. Light Van Selection	78
6. The Initial Effects of the Gross Weight Increase	91
7. The World Truck Concept	109
8. Research and Development Facilities	121
9. Road Testing Commercial Vehicles	137
10. The United Kingdom Bus and Coach Market	147
11. Keeping Wheels Secure	159
12. Commercial Vehicle Driving Techniques	167
13. Down Licensing and Down Plating	179
Appendix: Vehicle Weights and Dimensions	185
Part 2: Directory of Manufacturers, Distributors and Trade Associations	195
Index	227
Index of Advertisers	232

BENT AXIS....

what's all this about?

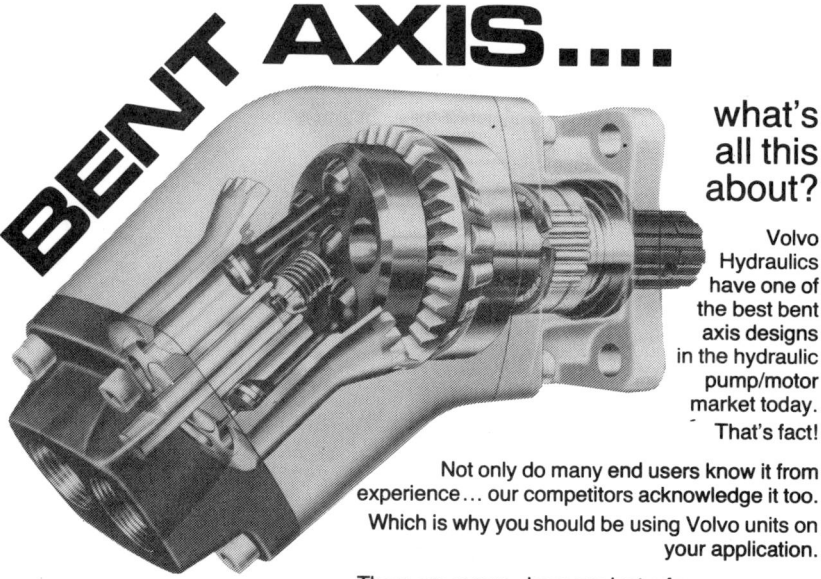

Volvo Hydraulics have one of the best bent axis designs in the hydraulic pump/motor market today.
That's fact!

Not only do many end users know it from experience... our competitors acknowledge it too.

Which is why you should be using Volvo units on your application.

There are many – here are just a few.

TIPPING TRUCK P.T.O. DRIVES

The lightweight F1 truck pump (top) has a choice of p.t.o's (some are shown above) to facilitate mounting on ZF, Fuller or Volvo gearboxes; this means you can use the pumps on most trucks of European manufacture including Volvo, Scania, Mercedes, M.A.N. etc. High power/weight ratios in all 6 sizes, e.g. 20 kW (6.7 kg – 66 kW (18 kg). Outputs from 20-110 cc/rev. at up to 250 bar continuous. Splitter Box (SB154) available for dual pump mounting to single drive.

POWDER OR GRANULE BLOWERS

The highly reliable F11 pump/motor (illustrated is the recently introduced 190 kW version) is an ideal unit for both ends of a hydrostatic transmission system; numerous are now being used to drive blowers or compressors as well as on-board winches, cranes etc.

Want any further information? Call, ring or write:
VOLVO HYDRAULICS LIMITED
130, Thornes Lane, Wakefield WF2 7TG
Tel: 0924 361616   Telex: 556193
Agents and Distributors throughout the country.

Introduction

One of the major influences on how a fleet engineer runs his business is that of legislation. Whether the fleet consists of one vehicle or one thousand the law, and the need to comply with that law, remains the same. Environmental pressure has increased considerably in recent years, especially in such areas as vehicle noise and exhaust emissions. The move to 38 tonnes gross weight in the UK incorporated many environmental suggestions for the future in addition to the confirmed legislation for such areas as side guards and rear under-run protection.

In the 1984 edition of *The Transport Engineer's Handbook,* a great deal of space has been devoted to legislation and to some of the ways with which to comply with that legislation. For example, vehicles built for the carriage of petroleum spirit must comply with certain stringent safety requirements which are fully explained in Chapter 3 of this book.

Reducing vehicle noise is a problem for the manufacturer rather than for the operator but if the trend towards quieter vehicles continues then engine encapsulation might be necessary, with its ensuing problems of maintenance difficulties and increased operating temperatures. The various factors influencing noise measurement and an explanation of the testing procedure are given in Chapter 4.

Since 1 May 1983, it has been possible to run at 38 tonnes gcw in the UK. This has required a minimum of five axles with the position of this extra axle being open to interpretation. The *Handbook* looks at the advantages and disadvantages of the various options and looks at some typical conversion costs.

The technical problems of Down Plating because of official refusal to allow Down Licensing are explained in Chapter 13. The law does not allow Down Licensing in spite of the fact that, to most people, it is a more logical solution to the problem of taxation than the technically irritating methods used in Down Plating.

Most road transport engineers rely heavily on Press road tests for making up a vehicle selection short list. But how does the trade Press carry out such tests and will these tests be superseded by computer? The current methods used by the Press are explained, as is the latest computer-based technology aimed at shortening the time taken to evaluate various technical options.

The term 'World truck' has received great publicity recently in the wake of various car manufacturers' comments on the concept of the 'World car'. But is the comparison a valid one for the commercial vehicle industry? It is often very misleading to equate the lorry with the car, this subject is discussed in detail in Chapter 7.

The vehicle and component manufacturers spend vast sums of money each year in developing new, and hopefully, improved, models. How they test them is explained in Chapter 8, with some of the end products being described in detail in Chapter 1. The latter includes a technical description of a continuously variable transmission as well as an up-date on the use of composite materials in the form of propellor shafts and leaf springs as used in Ford's Concept Cargo.

The light van tends to be a neglected animal in many fleets. Too often the initial selection is incorrect with too little thought going into it in comparison with the case of the top weight tractive unit. However, the correct selection of a light van is an important item in reducing fleet costs. Chapter 5 of the *Handbook* aims to assist in this process of selection.

After some time in the design doldrums, the passenger vehicle industry is now once again demonstrating some interesting technical developments especially in the field of integral construction. The 1984 edition of *The Transport Engineer's Handbook* looks at the current trends in psv design.

Aggravated by increased cornering speeds and more widespread use of power steering, the problems of wheel security are increasing. The subject is a complicated one as faults can show up in many different areas – for example, loose nuts, fractured studs and cracked wheels. Chapter 11 examines various aspects of wheel security along with some of the solutions to this perennial problem.

The winter of 1983-84 was a relatively mild one and so there were few instances of vehicles stranded because of 'frozen' diesel fuel. This is not always the case, however, as mentioned in Chapter 2, which explains just what can happen to derv in sub-zero climates if the right precautions are not taken. Whether a fuel additive is used or whether a fuel heater is fitted to the vehicle, some form of insurance against winter fuel problems is advisable and the various options open to the operator are covered in this, the 1984 *Transport Engineer's Handbook*.

Part 1

Why set up a profitable relationship with Eaton?

When it comes to truck components, you can't talk long life without talking Eaton.

Because the legendary reliability and durability of Eaton drivelines keep your vehicles on the road.

Because the servicing of Eaton drivelines is simple, needs no sophisticated tools and reduces downtime.

Because Eaton drivelines reduce driver fatigue through fast, easy shifting.

Because Eaton driveline parts are easily available through most European truck distributors and Eaton's own Sales and Service branches.

And because with six European manufacturing plants and a European Engineering Centre, Eaton are constantly researching, developing and innovating to produce better products.

No wonder Eaton driveline components are fitted to almost every major make of truck in Europe.

And this means that today, over one million trucks ride the road on Eaton and Fuller transmissions, Eaton single, 2-speed and tandem drive axles and Eaton brakes.

So next time, don't just specify driveline components, specify Eaton.

You simply can't buy better.

E·AT·ON
Truck Components

Eaton Limited, Truck Components Marketing, Eaton House, Staines Road, Hounslow, Middlesex. TW4 5DX Tel: 01-572 7313

1. Recent Technical Developments

Graham Montgomerie

The design process of a commercial vehicle never stands still even though, because of the whims of the legislators, the direction of this process might sometimes seem a little confused.

For example, there has been a great deal of interest in the transmission world in recent months with the announcement by Leyland and Scania of a continuously variable transmission and a computer aided gear changing system. The Leyland CVT is still very much in the prototype stage whereas the Scania concept is now in what is described as 'initial limited production'.

A continuously variable transmission has no distinct ratio steps, as is the case with a conventional manual or automatic gearbox. Although the basic principle has been worked on since the start of the motor industry it has never been developed to a point where a CVT was considered as a practical proposition for a production vehicle. Leyland claims that the key to its conception of a CVT has been the development of microprocessor-based electronic control equipment.

Historically, there have been three types of CVT: the belt type, used in the DAF (now Volvo) passenger car; the hydrostatic type; and the tractive type, with the latter being the path followed by Leyland.

In broad outline, the Leyland CVT consists of three main components: a variator, capable of giving an infinitely variable range within predetermined limits; an epicyclic section to provide reverse, neutral and forward; and the microprocessor-based controller.

The design of the variator unit is based upon an idea that can be traced back to at least 1899. It consists of sets of co-axial input and output discs which face one another, with the power being transmitted between them by further small discs or rollers arranged with their axes at right angles to the main shaft. These rollers effectively 'bridge' the gap between input and output discs in such a way that, if the input disc is turned clockwise, the rollers transmit the power to the output disc causing it to turn anti-clockwise.

There is no direct physical contact between the discs and the rollers; the drive is transmitted between them via an elastohydrodynamic oil film. Because the oil at the point of contact is subjected to enormous pressure – typically, 150 tons per square inch – a variation in its

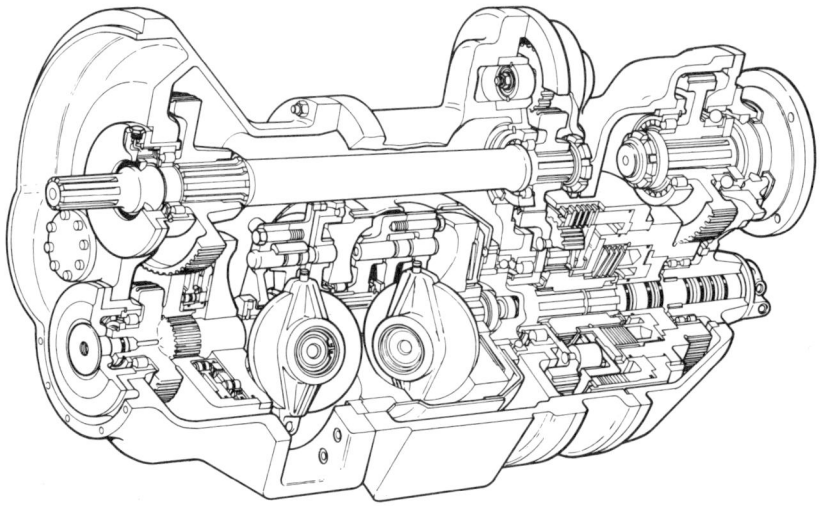

Figure 1.1 *The Leyland continuously variable transmission has no discrete ratio steps, as is the case with a conventional manual or automatic gearbox.*

physical properties occurs. In effect, the oil becomes semi-plastic causing a force to exist as the roller shears across the face of the disc.

The oil film causes total separation of the metal surfaces thus preventing erosion of the metal which, in theory at least, should mean no wear of the relevant surfaces. Although loosely referred to as oil it is perhaps better defined as tractive fluid, which is specially formulated and thus expensive. As it will not be degraded by flash debris or products of combustion, as is the case with conventional lubricating oils, the working life of the fluid should be a long one.

The faces of the input and output discs have semi-torroidal tracks and are so arranged that the inter-disc rollers can be moved so that they connect the outer edges of the input discs with the inner edges of the output discs or vice versa. The continuously variable part of the transmission is provided by the fact that the rollers can be positioned at any point between the minimum and maximum radius of the input and output discs.

The method of moving the inter-disc rollers, to vary the ratio, is by steering them which involves a castor angle, causing the variator to rotate only in one direction. Because of this, and the fact that the variator is permanently connected to the engine, reverse and neutral have to be provided by means of an epicyclic assembly where the Sun gear is driven by the variator and the planet carrier by the engine (see Figure 1.3). The output to the road wheels is taken from the annulus. By

Recent Technical Developments

Figure 1.2 *With the Leyland CVT the power is transmitted between the input and output discs by means of small rollers, two of which can be seen here.*

using the variator to create differing rotational speeds between the engine-driven planet carrier and the variator-driven Sun gear, the annulus can be driven forward or backward. It can also be held stationary, in a condition which is known as 'geared neutral'.

This also enables the entire transmission to operate in two phases: low regime, produced epicyclically, which is used for reverse and for forward speeds up to the equivalent of second gear, and high regime, where the drive is taken directly from the variator input.

Incorporated in this epicyclic unit (which owes a lot to Leyland's experience through Self Changing Gears) are annular clutches to hold the high or low regimes, peculiar to this type of clutch, which are equivalent to the brake bands in a conventional SCG gearbox.

The minicomputer for the CVT is packaged in the standard Leyland automatic transmission control box. It is, however, considerably more complex than this. The conventional automatic transmission controller can only make three decisions: to change up, to change down or to do nothing. Usually it is required to do nothing at all. The CVT controller, on the other hand, by the very definition of the term continuously variable is required to act continuously.

The unit is based on a 16-bit microprocessor and has 8K of control

The problem can be solved simply by contacting ROR and asking for details on our/ ROR Neway's AR series Air Suspensions.

Combine this new Air Suspension with the revolutionary TL 18000 series axle, designed to meet your 38 tonnes tri-axle requirements, and you will be running on the lightest most cost effective tri-axle configuration we know.

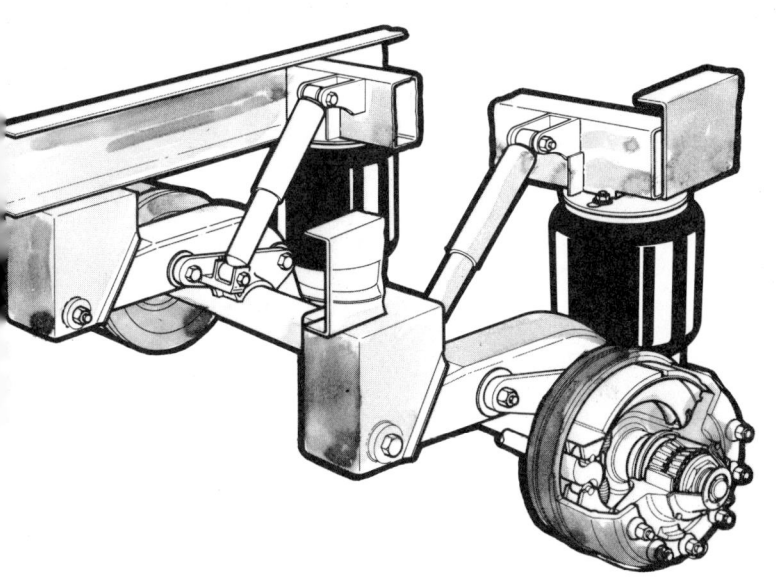

...nufacturing Plant: Rackery Lane, Llay, Nr. Wrexham, ...vyd LL12 0PB, Wales
...ephone: (097-883) 2141 Telex: 61427

Figure 1.3 *The faces of the input and output discs have semi-torroidal tracks (left). The epicyclic section (right) provides reverse and neutral.*

software and 2K of read-only memory. The program monitors road conditions and provides the control signals for the operation of the transmission.

The CVT in its present form has a power handling capability of around 375kW (500bhp) and it weighs 330kg (6.5cwt).

A further development scheduled for the Leyland CVT is that of regenerative braking which involves capturing, storing and using the energy currently wasted during the braking process. To illustrate the potential for such a system, consider the case of a typical 16 ton gvw 4 × 2 rigid. Such a vehicle is normally driven by an engine with a power output of between 150 and 250bhp. When such a vehicle makes an emergency stop at 0.5g from 30mph about 1,400bhp is instantaneously wasted in the form of heat.

Leyland is investigating the use of a high speed flywheel as an energy store and an efficient CVT is essential if such a system is to recapture

SHORFAST
LOAD SECURING EQUIPMENT

Full range of Cargo Straps, Ratchet Tensioners, Side Curtain Straps, Buckles, Horizontal and Vertical Wall Track, Slimline Floor Track, Garment Rails, Beam Brackets, Shoring Bars, Special Expanding Shoring Bars for use without track, Multi-trip Inflatable Dunnage Bags.

SHORFAST Ltd., Charles St., Chester, CH1 3EL, England.
Tel: 0244-49868. Telex: 61125

---

and use this wasted braking energy successfully. In this case, the flywheel would be accelerated by the decelerating road wheels during braking which demands a continuously changing transmission ratio.

Once spinning, the flywheel will store this energy which can be used to get the vehicle moving from rest and accelerate it up to around 15mph when the normal engine would take over again.

Leyland claims that the benefits of the combined CVT and regenerative braking system could be fuel savings of up to 30 per cent.

Initial development work at Leyland is being concentrated on the installation in a Leyland National bus, although the company also has a CVT operating in a 7.5 tonne Terrier chassis. Leyland claims that the potential for the CVT goes far beyond that for conventional road going vehicles to include railcars, off-highway construction equipment, agricultural machinery as well as tracked and all-wheel drive vehicles for both military and civil applications.

The Scania development is an automated rather than an automatic transmission as accepted in today's automotive vocabulary. Scania argues that the driving techniques necessary to exploit the potential economy of the latest generation of low-revving, high torque engines require a greater degree of concentration on the part of the driver than

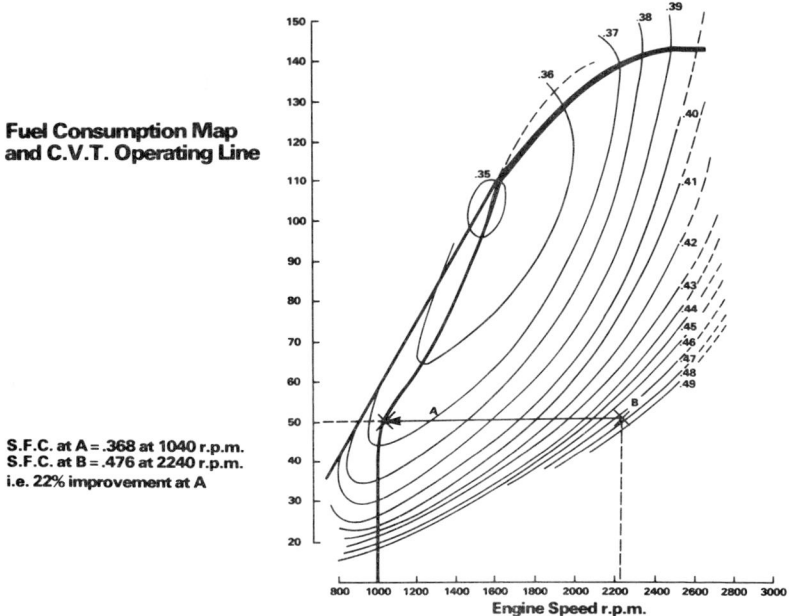

Figure 1.4 *The aim of the Leyland CVT is to keep the engine operating at or near its point of maximum efficiency.*

was the case in the past. This has stimulated interest in systems designed to ease the gear changing part of the driver's job.

As with the Leyland CVT, a major factor in the Scania system has been the rapid acceleration in microprocessor development for control units. Such a microprocessor is fitted in the cab and is programmed with all the parameters appropriate to the vehicle to which it is fitted. This also receives information from vehicle status monitoring units including road speed, changes in relative road speed, throttle position and the gear which is engaged. The control unit then works out which gear should be used, informing the driver by means of a light on the instrument panel.

Under most circumstances when the driver sees that another gear is needed, he merely has to press the clutch pedal in the normal way and air-operated servos, mounted on a virtually standard Scania ten-speed synchromesh gearbox, would then carry out the actual selection of an appropriate gear. This process is claimed by Scania to be faster than normal manual shifting.

With this automatic pre-selection system the control unit selects the optimum gear on the basis of the information available to it. However, it cannot see if the vehicle is approaching a steep gradient and thus an

Transport Engineer's Handbook

Figure 1.5 *In its present form, the Leyland transmission has a power handling capability of around 500 horsepower.*

override facility is incorporated. For example, a situation could arise where a vehicle is proceeding along a flat road in ninth gear and the tenth gear light comes on. If the road continued to remain as a level surface then the driver would merely depress the clutch for tenth gear to be engaged. However, if the driver can see that he is approaching a hill he might want to change down rather than up. To achieve this he has to move a toggle switch positioned where the gear lever would normally be in a conventional vehicle. By pulling the toggle switch back the computer is then programmed to drop one gear which is then engaged when the driver operates the clutch.

Gear changing by means of this toggle lever works on roughly the same principle as that used on motorcycles. Each forward movement of the switch programs an upward change while each movement to the rear changes down. Two pulls on the lever would result in two lower gears being chosen and so on, again with the actual ratio change being made only when the clutch is operated.

A selector switch is located alongside the miniature gear lever by which the driver can select reverse, neutral, manual or automatic preselection.

Various safeguards have been incorporated into the Scania system including a device to prevent the engine from over-revving if too low a gear is engaged. A further safety feature prevents the engagement of a low gear if the wheels are locked or if a speed transducer signal failure

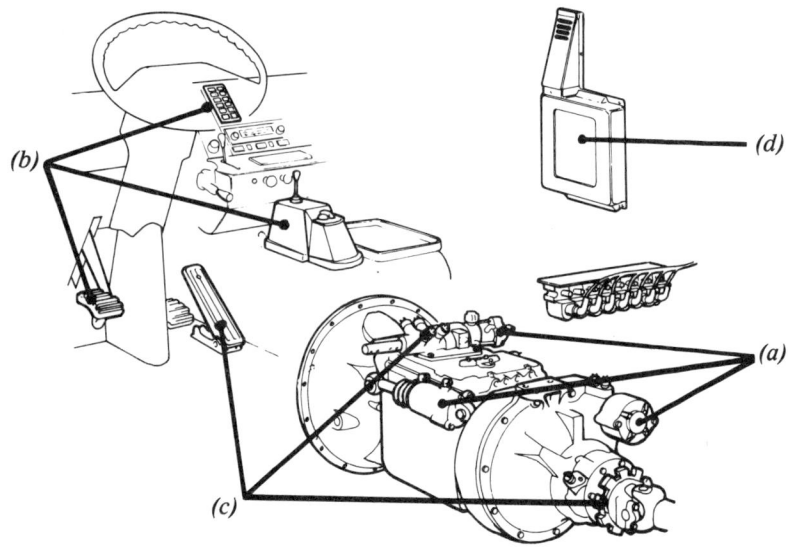

Figure 1.6 *The Scania computer-aided gear changing system. (a) Actuating cylinders to change gear. (b) Driver communications unit: gear shift indicator, stalk, mode selector switch and clutch pedal with buzzer to tell the driver when the gears are being changed. (c) Vehicle status monitoring units: throttle, gear in use and road speed. (d) Processor and control electronics to assess vehicle status parameters and recommend suitable gears.*

occurs. If there is a total failure of the automatic equipment then a 'standby stick' is provided to change gear manually.

A built-in test program is incorporated to allow the driver or the workshop fitter to check out the function of the various signal transducers and to pinpoint the cause of any failure.

Unlike hydraulic automatic transmissions that change gear without loss of torque, existing designs of mechanical gearboxes need the torque to be disconnected via the clutch before the ratio change can be made. Scania suggests that in the case of multi-speed (ten or more gears) close ratio gearboxes, such torque losses would be highly inconvenient if efforts were made to automate the gear changing process. This is the reason that Scania has given for allowing the driver to exercise his discretion in determining when the traffic situation allows him to accept such torque losses by having the actual physical gear change activated by the clutch pedal.

Although every vehicle manufacturer worldwide has its own research and development operation (some large – some not so large) it is rare for the factories to release details of current research projects. Ford went a few steps further than merely stating which lines of thought the

Recent Technical Developments

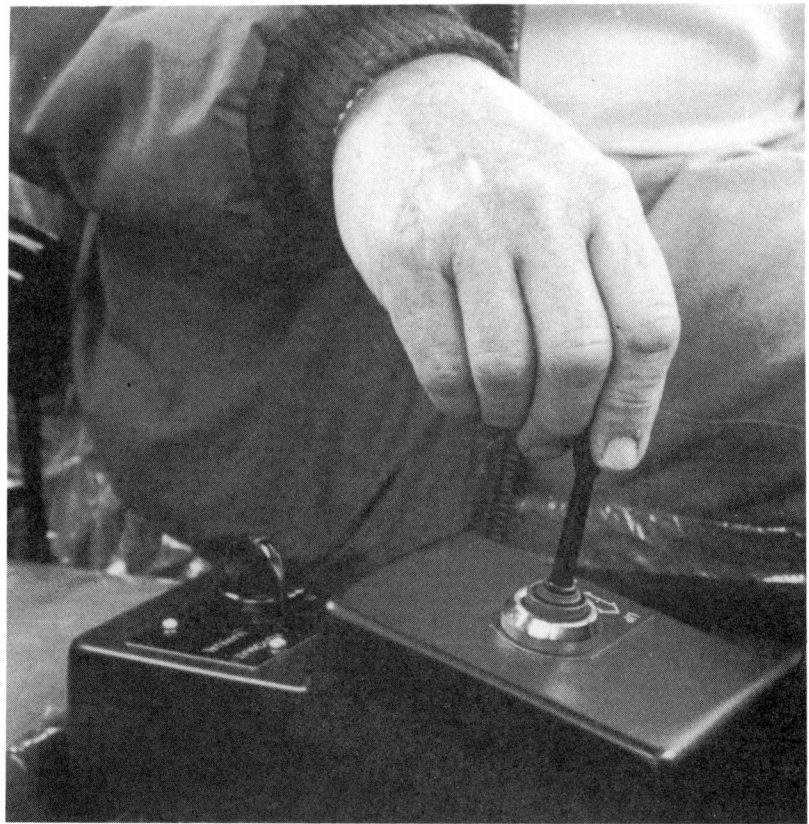

Figure 1.7 *Using the conveniently located shift stalk, the driver can select a gear other than that recommended by the computer. He uses the switch behind the lever to select automatic or manual mode.*

company was following by producing a concept vehicle which incorporated a number of experimental features. Known as Concept Cargo it was built as a 'one-off' to show the world what Ford was trying to achieve.

Concept Cargo was never intended to be a prototype for a production model; it was intended more as a mobile test bed to evaluate a number of features which may be incorporated into vehicles of the future. The overall aim of Concept Cargo was that of improved operating economy with the emphasis very much on *overall* operating economy; it did not refer merely to fuel economy for instance.

The Ford development team grouped their targets into four basic categories:

Figure 1.8 *For Concept Cargo, Ford worked with Windfoil to produce a light-weight van body with good aerodynamics.*

○ improved aerodynamics
○ improved engine efficiency
○ optimised drive-line matching
○ reduced kerb weight

For Concept Cargo, Ford worked with the Norfolk-based body builder Windfoil to produce a lightweight van body with good aerodynamics. The frontal area was minimised and the overall drag reduced almost to the minimum possible for this type of vehicle using a combination of aerodynamic aids. Windfoil incorporated a roof-mounted deflector, an under bumper air dam, side and rear skirts and also 'streamlined' the rear bodywork.

The floor of the body was constructed of 50mm (nearly 2in) thick Ciba-Geigy Aerolam aluminium honeycomb with an epoxy-resin skin and incorporated integral runners and cross bearers attached using an epoxy-resin adhesive. The runners sat directly on the chassis being attached with conventional Cargo body mountings.

The 18mm (0.7in) thick side panels had hot-pressed polyester and woven glass skins around a core of end-grain balsa to give a high impact resistance combined with a low weight of only 6.8kg/m² (1.4lb/ft). An epoxy-resin adhesive was again used to bond the sides to the floor.

By extending the sides some 50mm beyond the bottom of the floor, additional stiffness was added while at the same time eliminating the need for a side rave.

The Ford engineers developed the standard adjustable air deflector into a fixed fairing specially constructed for the Windfoil body fitted to Concept Cargo. Normally there is a considerable gap between the cab and the bodywork, and wind tunnel work has shown that this gap can upset the smooth flow of air over the sides and top of the box.

In the Concept Cargo design, the air flow into the gap was eliminated by the addition of a collar which extended the natural surface of the roof deflector and the cab's B pillars. The collar was hollow and featured a duct at cab roof level to feed air to the conventional cab-mounted engine air stack.

When the cab was tilted the roof fairing moved up with the cab, whereas the collar remained fixed to the box.

The rear of the body was finished off with a D-shaped moulding which was designed to reduce the area of the vehicle's turbulent wake and thereby reduce the overall drag coefficient.

The weight of the complete body including the extra roof fairing, collar, under bumper air dam and skirts was 686kg (13.5cwt) which was comparable to an equivalent aluminium body, in conjunction with the currently available Ford-designed roof deflector, and some 200kg (4cwt) lighter than an equivalent body using a glass fibre/plywood composite construction – again with the optional Ford roof deflector.

For the cab floor of Concept Cargo, Ford used a sound-deadened steel from Antiphon which consists of cold rolled steel facings with a visco-elastic interlay. This sandwich is lighter than steel of the same nominal thickness and can be formed using conventional press tooling. Although Antiphon is relatively expensive, its selective use could save weight by reducing other insulation – typically, a thick rubber/foam/rubber sandwich mat, currently used to reduce in-cab noise, can weigh up to 50kg (110lb).

A term which is becoming increasingly familiar in the commercial vehicle world is 'alternative materials' which is used to describe anything not considered the norm for a production component. Perhaps carbon fibre is the best example of such a material and, even though it is used in many different forms, it is still referred to as carbon fibre in the singular. On Concept Cargo, the science of alternative materials was applied to the prop shaft which, on the production Cargo, is a two piece assembly. The material used was a fibre-reinforced araldite epoxy/

Figure 1.9 *Glass fibre composite leaf springs were fitted to Concept Cargo. The disc brake assembly used Lucas Girling reaction beam hydraulic calipers in conjunction with Ford-designed hubs and discs.*

carbon fibre composite from Ciba-Geigy which has similar strength characteristics to steel but with the advantage of better internal damping to isolate axle vibrations.

On the 3.725m (12ft 3in) wheelbase Concept Cargo, the use of this material allowed the two-piece prop shaft to be replaced by a single shaft which saved 6kg (13lb) in weight.

Ford engineers claim that lightweight leaf springs made from glass fibre and resin composite could one day take over from minimum leaf springs in the way that minimum leaf springs are replacing multi-leaf assemblies. The main attraction of such composite springs is their lightness; they are approximately one-third of the weight of an equivalent multi-leaf spring and less than half the weight of a minimum leaf design.

The springs fitted to Concept Cargo were developed to perform the same function as the standard Ford minimum leaf springs and achieved a weight saving of 87kg (190lb) – equivalent to 60 per cent.

A further advantage is that glass fibre composite springs have inherently good noise damping characteristics and this reduces the transmission of axle noise to the chassis.

GKN manufactured the composite springs for Concept Cargo with the company claiming that costs are potentially comparable to steel and

Recent Technical Developments

Figure 1.10 *The disc brake system saved 19kg (42lb) compared with an equivalent drum arrangement and halved the parts count.*

are likely to become even more competitive in the future. This statement contradicted most of the comments put forward in the past about production costs for composite materials, but the argument was based on the fact that composites require only half as much energy to produce as steel and because the production of the components from the composite raw material costs only one third to one fifth of that required to produce steel springs from steel stock. GKN also argued that this potential cost saving was likely to increase in direct line with increases in the cost of energy.

Although disc brakes have been used for over 25 years in production vehicles, their use has been confined to passenger cars and goods vehicles below 3.5 tonnes gross. Concept Cargo was fitted with disc brakes all round in contrast to most cars and light commercials, which tend to favour discs at the front only while retaining drums at the rear. The Ford choice was because even under very heavy braking, with its resultant load transfer, up to half the braking force is still provided by the rear axle.

The claimed advantages of disc brakes include consistency of performance, a progressive response with excellent stability over a wide range of operating temperatures, lower weight, compared to a drum assembly, a reduced parts inventory and lower pad replacement costs. The reasons that such claimed advantages have not resulted in universal acceptance are that, in common with most engineering solutions, there area number of disadvantages with disc brakes. Parking brake efficiency is one problem area, while another is finding the space within

Figure 1.11 *Concept Cargo was used as a mobile test bed. Note the lattice construction of the Mathweb side guards.*

the wheel to provide sufficient frictional area. With a drum brake, extra lining area can be found by increasing the width of the drum. The equivalent increase is not possible with a disc brake housed as it is within the confines of the wheel. For this latter reason, high disc and pad wear is still considered to be a problem.

The disc brake assembly fitted to Concept Cargo used Lucas Girling reaction beam hydraulic calipers in conjunction with Ford-designed hubs and discs. As fitted, the system saved 19kg (42lb) in total over the drum brake layout and reduced the parts count by half, thus vindicating at least two of the claims made for disc brakes in general.

To achieve the required parking brake performance, Ford

Figure 1.12 *Motor Panels produced this Leyland C40 cab in which some of the standard steel pressings were replaced by components in other materials.*

incorporated a double-acting thread mechanism to provide automatic adjustment. The system was operated by specially developed Bendix air brake actuators connected directly to the calipers. The brakes themselves used an air/hydraulic system modified to increase fluid delivery and capacity.

In line with current environmental pressure, DON 7250, a new semi-metallic asbestos-free friction material, was used for both front and rear pads. Criticism levelled at non-asbestos linings includes high wear and high cost, but Ford has claimed that development trials have indicated an acceptable life with the DON material.

Most production Cargo chassis are fitted with power steering as standard equipment but the Concept Cargo exercise went one stage further

Figure 1.13 *These six panels from the Roadtrain cab were produced on the same tooling but from different materials.*

and incorporated a proportional feel system. For low speed manoeuvring or parking, full power assistance was available with this assistance gradually decreasing as vehicle speed increased.

In a power steering system fitted with a rotary control valve, as in Concept Cargo, the amount of power assistance is determined by the stiffness of a small torsion bar. The TRW/Cam Gears system developed for this particular vehicle had a double ball and cam arrangement fitted to one end of this torsion bar. Hydraulic pressure could be applied to these balls thus increasing the apparent stiffness of the torsion bar and hence reducing the amount of power assistance. The hydraulic pressure was provided by a pump driven by the propshaft thus providing the proportionality to the vehicle speed.

In most respects, the engine for Concept Cargo was the current production 6.2 litre six-cylinder unit. However, it was modified in

CURL UP WITH

In 1981, Ford woke up the entire transport business with the launch of the Cargo.

Within seven months, it became Britain's best-selling truck.

Now there's a factory-built sleeper cab option on this range.

Unbeatable flexibility.

Not only does this make Cargo the only truck offering a sleeper cab on all gross vehicle weights between 7.5 and 32.5 tonnes.

It does so with a complementary range of engines, from 110 to 204 bhp.

If you have to take loads over long distances, Cargo's the only reasonable choice.

More features as standard.

Like all '84 Cargos, the sleeper cab mod have more features as standard.

There's a stylish, comfortable bunk with a fully adjustable roof hatch to give you the b possible sleeping conditions.

A single passenger seat is standard. And s everything you need to fit an audio system: wiri aerial and two built-in speakers.

More facilities for the driver.

Cab comfort is one of the prime conside tions on the award-winning Cargo.

The driver's seat is adjustable for height, reach, cushion angle and rake.

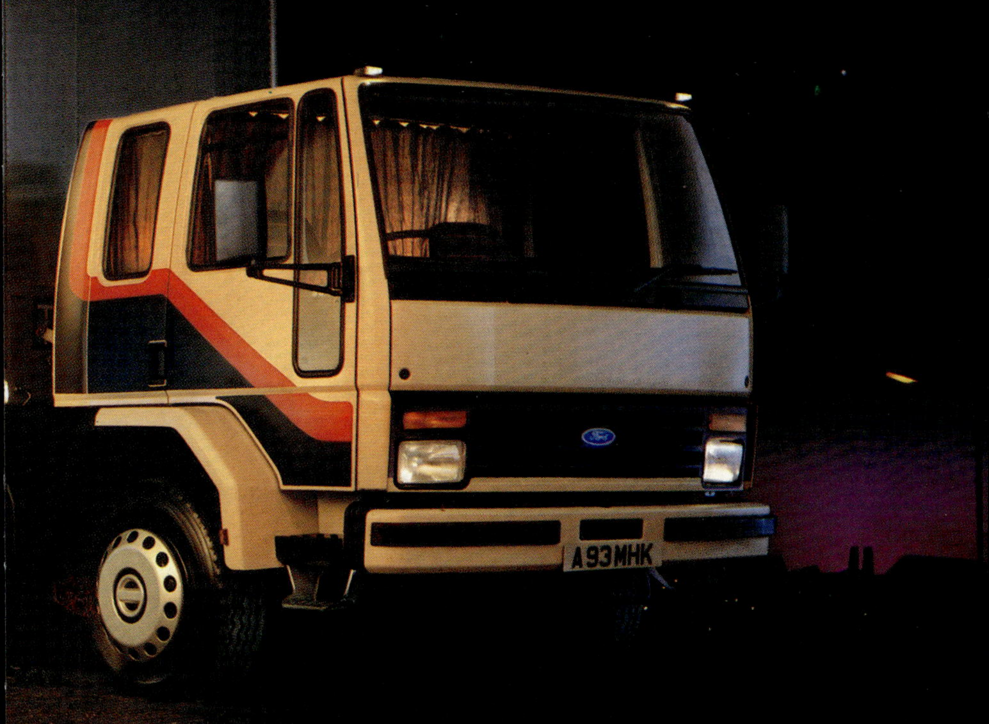

A BEST SELLER.

Deep, kerbside observation windows give additional visibility where it counts. Vital features that, on a Cargo, won't cost you a fortune.

Less drag. Longer service intervals.

Cargo's remarkably low drag coefficient gives you greater energy saving than ever before.

The drag-link, gear-change mechanism and steering joints are lubricated for life, and the clutch is self-adjusting.

Ford gives you more.

As if all this wasn't enough, every Cargo is now available with Truck Extra Cover, a second year optional warranty plan on major driveline and steering components.

And every Cargo is backed by Britain's finest national network of Truck Specialist Dealers.

Talk to your local Ford Truck Specialist Dealer about the sleeper cab version of the award-winning Cargo. He'll tell you about the great deals he can offer.

FORD CARGO
6-32·5 TONNES

BRITAIN'S BEST SELLING TRUCK.

J.W.E. Banks & Sons Limited

IMPORTERS AND DISTRIBUTORS OF AUTOMOTIVE PARTS

Koni's – the widest range of dampers in the world – cars, buses, formula 1, commercial, agricultural, motor cycles, railways, industrial, engine dampers, steering dampers, even bridge dampers.

Contact the sole U.K. concessionaires J. W. E. Banks & Sons Ltd. when you need specialized adjustable dampers of the very highest quality.

Crowland, Peterborough PE6 OJP. Tel: (0733) 210316 Telex: 32533.

Sole U.K. Concessionaires   Koni Adjustable Shock Absorbers.

certain major areas. The rated speed was reduced by 23 per cent from 2,600 to 2,000rpm and the maximum torque speed was lowered from 1,500 to 1,200rpm. To compensate for the inevitable loss in power inherent with a large reduction in engine speed a turbocharger was added to retain the power output of 82kW (110bhp).

As well as maintaining the power level, the use of the turbocharger increased maximum torque by 31 per cent from 335 to 440Nm (248 to 325lb/ft).

The lower engine operating speeds were employed to reduce internal friction and parasitic losses within the engine, thus improving mechanical efficiency. Together with the improved thermal efficiency afforded by the turbocharger, this resulted in a predicted 12 per cent improvement in fuel consumption according to Ford computer studies.

Although the turbocharger added weight to the engine, this could be offset partly by savings in cab noise insulating material because of inherently lower noise level of turbocharged engines.

A 'clean hands' method of checking the oil level was fitted to Concept Cargo in the form of a fibre optic sensor which relies on a beam of light from a light emitting diode at the top of the dipstick being internally reflected from the base, with the returning signal being used to trigger a suitable warning lamp. If the base of the dipstick is below the oil level, the light is not reflected and the lamp remains off.

On Concept Cargo, the system was wired through the ignition key and the oil level was checked just prior to cranking; it did not display while the engine was running.

One of the areas of operating economy not directly related to increased fuel economy or potential payload is the avoidance of overloading fines, while at the same time making full use of the vehicle's carrying capacity. There have been many instances where operators have run their vehicles below their gross weight capability merely to eliminate any possibility of being caught by a roadside check. Where a weighbridge is on-site this is not a problem, but in many cases one is not available so the vehicle is loaded by a combination of experience and guesswork.

On-board weighing devices were developed to solve the two interrelated problems of overloading and underloading and Concept Cargo was fitted with such a device. The Loadax system, manufactured by TRW Probe, consists of three strain gauge transducers, one of which, on Cargo, was fitted on top of the front axle beam with the other two beneath the rear axle and mounted on either side of the axle bowl. As the load on the axle increases, the flexing of the transducer produces a signal which is directly proportional to the load producing it.

With Concept Cargo this signal was fed to a digital read-out in the cab where it was claimed that the readings were accurate to the nearest 100kg or better than 2 per cent. The Loadax could be adjusted to read gross or net weights and to give a warning if the load should move as a result of cornering, accelerating or braking and overloading either axle.

All production Ford Cargoes are equipped with steel-belted radial-ply tyres with sections of between 80 and 85 per cent, but Concept Cargo ran on Dunlop 75 per cent section low profile tyres. These squatter 225/75R 17.5 tyres have a smaller static radius than the conventional 9.5R 17.5 radials and reduced the loading height of Concept Cargo by some 30mm (1.2in).

Concept Cargo was fitted with a conventional rear guard to meet EEC 78/1964 requirements but the side guards were of an interesting construction to meet the proposed legislation for rigid vehicles. The beams were built by Bristol Composite Materials from Mathweb, a patented form of lattice construction. The pairs of beams weighed only 15kg (33lb), about one third the weight of an equivalent steel beam.

The widespread use of the tachograph and the recent advances in microcomputers have been combined in Concept Cargo with the incorporation of a data collection device developed by Vehicle Control Systems. Known as VEDAC, it is a small module incorporating a microprocessor measuring $145 \times 90 \times 30$mm ($5.7 \times 3.5 \times 1.2$in) which in the case of Concept Cargo was mounted next to the driver between the seats. Into this module was fitted a $76 \times 62 \times 20$mm ($3.0 \times 2.4 \times 8$in) cartridge which is removed and connected to a microcomputer when an analysis is required.

Although the main purpose of VEDAC, as installed in Concept Cargo, was to store all the information recorded by the tachograph, the basic system could also be used to collect up to five additional sets of data. For example, with a suitable transducer installed, VEDAC could monitor all fuel added or taken from the vehicle's fuel tank and record all fuel used by the engine in relation to distance and time.

The search to find better materials with which to do a particular job has been highlighted again by an exercise carried out by Motor Panels (Coventry) Ltd. This took the form of a Leyland C40 cab in which some of the standard steel pressings were replaced by components in other materials in the search for lighter weight and improved sound deadening characteristics.

The first stage of the exercise started back in 1982 at the Birmingham motor show when Motor Panels displayed six examples of the upper front panel of a Roadtrain cab (the panel carrying the Leyland name), all produced in a different material but using the same tooling as for the standard component.

The materials used were mild steel, zinc-coated steel (Zintec), stainless steel, Lite-Plate (since renamed Steel-O-Bond), aluminium and Azdel. The aluminium was the lightest at 2.88kg (6.35lb) which compared with the 6.83kg (15.1lb) of the heaviest panels which were the mild steel and the Zintec.

Steel-O-Bond is a steel/propylene/steel sandwich with the steel layers being only three thousandths of an inch thick. Its main advantage is its excellent ability to dampen down sound but this is counterbalanced by the fact that it cannot be spot welded. It is however perfectly satisfactory where it can be bonded.

The Lite-Plate (Steel-O-Bond) panel shown at Birmingham weighed 3.32kg (7.32lb) compared to the 4.03kg (8.86lb) of the Azdel which is a polypropylene material reinforced with glass fibre. This comes in thin sheets which are then heated to 205°C (400°F) causing them to double in thickness. The sheet is then pressed between the forming dies.

Having proved the point, the next stage was to incorporate some of these materials into other areas of the cab, the floor in particular. The cab floor has a long list of special requirements as far as the designer is concerned. It must be structurally strong and corrosion free as well as acting as a seal against noise. If this were not enough it has to be able to cope with a variety of shapes and sizes of engine and to be easy to clean.

Thus the Motor Panels' prototype is of a four-layer sandwich construction plus a 'bonded-on' top surface which eliminates the need for floor mats. The first and third layers are of dense polyurethane foam while layers two and four are glass-reinforced polyester with Molochite filler. This composite floor is about the same weight as current designs in more conventional materials but it is claimed to be some 15 to 20 per cent better at reducing transmitted noise.

The outer door panels were made from Steel-O-Bond which is marginally heavier than aluminium yet lighter than steel. The quarter panels of the cab were in aluminium alloy as was the front grille panel although the actual alloy was not identical for ease of production reasons.

The inner door panels were in seven microns thick double-sided zinc-coated steel which is the standard material for this C40 cab component, whereas the bulkhead was a high strength steel from British Steel known as Tenform.

The Leyland cab was used for this particular exercise as it was designed from the outset with the long-term use of alternative materials in mind.

Motor Panels claim that such materials, for specific components, could be incorporated into current production cabs within a year, although for a new cab this would take the same length of design time as for any other new cab and that could be anything up to five years.

2. Diesel Waxing and its Prevention

Graham Montgomerie

In the severe winter (by UK standards) of 1981-82, scenes of drivers lighting fires beneath their vehicles became quite a common sight on roads the length and breadth of the country. This was not an example of final frustration with the manufacturers' after sales service, but a desperate attempt to unfreeze the fuel in the tank.

In point of fact diesel does not 'freeze' as such, it waxes up – although the net result is the same; the vehicle will not start because of fuel starvation to the injectors.

Derv is not one substance but a complex mixture of hydrocarbons which all have a different boiling point. As these boiling points can vary between 170°C to 380°C (340°F to 720°F) it is clear that they will not all 'freeze' at the same temperature.

The paraffin hydrocarbons in diesel fuel have only a limited solubility which means that they tend to precipitate out in the form of minute wax crystals which remain in suspension in the fuel. It is these crystals which cause the blockages in the fuel system erroneously attributed to the diesel 'freezing'.

Although the viscosity of fuel increases with decreasing temperature this is not the limiting factor; fuel starvation, because of the accumulation of wax, occurs at a higher temperature than that at which the viscosity starts to become noticeable.

The colder the ambient temperature, the greater the tendency for these paraffin hydrocarbons to precipitate out as wax crystals.

The temperature at which these crystals start to precipitate out is known as the cloud point. In extreme cases the temperature can sink to a point where not only is there crystalline wax in the system but it is alsopresent in sufficient quantities physically to prevent the fuel from flowing – the temperature at which this happens is known as the pour point.

Neither of these temperatures, however, can be relied upon to give a precise indication as to how a fuel will perform in service as, in most cases, the cloud point will give a pessimistic reading in contrast to the pour point which gives an over-optimistic estimate. To get over this problem a test was developed to isolate a specific temperature which could then be used with an acceptable degree of accuracy to define the

37

Figure 2.1 *Conditions like these are not confined to Northern Canada where this picture was taken. Well-below-zero temperatures also occur frequently in the UK.*

low temperature performance of the fuel. This test is known as the cold filter plugging point test or CFPP for short.

In this test a measured quantity of fuel is drawn through a fine mesh screen. The temperature is gradually lowered and the test repeated until the flow of fuel through the screen is considered inadequate. The temperature at which this occurs is defined as the cold filter plugging point for that particular grade of fuel.

Diesel fuel sold in the UK is required to conform to BS 2869 which specifies a maximum CFPP of -9°C (16°F) during the winter period and a maximum of 0°C (32°F) for the rest of the year. It is difficult, however, to define accurately the term 'winter period'.

Production of winter grade diesel starts well in advance of the specified period, according to the oil companies, so that in theory the bulk tanks in their distribution network should contain the right grade by the time it is needed. This 'grade' of fuel suitable for winter use is obtained by incorporating additives which modify the structure of the wax crystals as the temperature drops. The additives do not prevent the waxing but they do ensure that the crystals are much smaller and thus less likely to clog up the system.

Diesel Waxing and its Prevention

The reason why the oil companies do not incorporate such additives throughout the year is purely one of cost. If there was a definite starting point to the winter then both operators and oil companies would know where they stood, but the winter of 1978-79 with its snow and ice lasting until mid-April is a good example of how impractical it is to attempt even to define 'winter period'.

Thus, although the oil companies try to ensure that winter grade derv is in the main tanks before the freeze starts, it is all too easy for the operator still to have summer grade fuel in his own tanks at the beginning of winter. In the spring when the oil companies revert to summer fuel, a late cold snap can bring about the same problems just when everyone is looking forward to a trouble free summer.

It can be seen from the foregoing that relying merely on the oil companies and an accurate weather forecast could be considered as foolhardy. It makes sense then for the operator to take as many precautions as he can to prevent his fleet being stranded.

The larger operator can insist that his fuel supplier will have the right fuel available at the right time, but this is easier said than done and is not a great deal of help to the owner driver who does not enjoy the bulk buying power of the big fleet.

For the fleet with its own depot storage tank this is a good place to

Figure 2.2 *Engine manufacturers spend a lot of time in testing their products at extreme temperatures, but the engine will only start if the fuel can reach the injectors.*

start and the first principle is to locate the tank in a sheltered position protected from the direct force of the wind. The pipework should be of large diameter with no sharp bends or other constrictions and should be well lagged with a water-proofed insulation.

Water and debris invariably accumulate in the bulk tank during the year and provision should be made for this to be drawn off. Installing the tank at a slight angle means that all the sludge will settle at one end where it can be removed via a drain valve. It is also important to ensure that the fuel pick-up point is situated well above the bottom of the tank, again to minimise the risk of drawing off the sludge. Complete draining of the tank just before the onset of winter is recommended but it has to be admitted that this is not always practical.

If strainers or filters are fitted near to the tank then these should be of a reasonably large size and of a fairly coarse mesh. In this way the debris is collected but without the major risk of clogging up with wax if a fine mesh were to be used.

Moving on to the vehicles themselves, overnight under-cover storage is rarely a practicable move but a good compromise is to park the

OCTAGON RECOVERY

SPECIALIST RECOVERY SERVICE

Roadside assistance anywhere in mainland Great Britain twenty four hours everyday, three hundred and sixty five days a year.

Commercial vehicles and fleet cars only.

SCHEME 'A'
Nominal Enrolment fee per fleet; £200 roadside credit; fixed rates for assistance; each job invoiced individually in arrears.

SCHEME 'B'
Annual Subscription per vehicle to cover free:-
1. Roadside assistance and recovery;
2. Repairs in workshops after recovery;
3. Extended Warranty on major components.

**Octagon Recovery Limited,
Tithe House, Town Street,
Horsforth, LEEDS LS18 5LJ.
Tel: LEEDS (0532) 583144.**

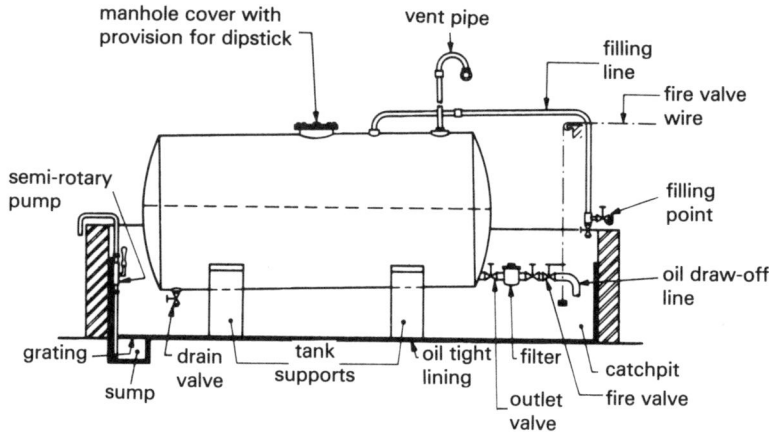

Figure 2.3 *This diagram demonstrates the principles of good installation for a depot storage tank.*

vehicles nose-in to a wall – or to one another – to minimise the effect of a high wind at zero degrees on the fuel lines and cooling system. In this context any form of temporary windbreak is to be recommended.

Lagging of the vehicle fuel tanks is a cumbersome method of protection but if a very severe period of cold weather is forecast then it is well worth the effort. Lagging the fuel lines is far easier and, because of the high risk of waxing in these areas, should be included in the 'must do' category.

In exceptionally cold spells, having the engines idling might seem an expensive method of protection but the extra cost in fuel must be balanced against the potential loss in revenue from a fleet which would not start when there were ferries to catch and loads to deliver.

Most vehicle manufacturers are well aware of the problems of cold weather operation but nevertheless there have been a number of instances of diesel waxing because of totally unsuitable chassis installations. To provide advice on the subject, the British Technical Council of the Motor and Petroleum Industries has published guidelines for fuel system layouts which are illustrated in Figures 2.4 and 2.5. Devised by the BTC project group on low temperature operation of diesel fuel system for vehicles – the group includes members from most of the oil companies and the vehicle and engine manufacturers – the guidelines suggest various 'do's and don'ts' for chassis installations.

For example, the fuel tank should be installed in a sheltered position and the location of the fuel feed and return pipes should be such that, under cold start-up conditions, a reservoir of warmed fuel is established quickly in order to keep the wax in solution.

The sizes of all the fuel delivery lines should be in accordance with the engine makers' requirements, which are reasonably easy to achieve with a new vehicle, but, once a few service schedules have passed, then anything is possible. The aim of a good fuel system design is to provide the minimum restriction to the fuel flow, which means that the bends in the system should be of the maximum practicable radius.

Fuel filters and agglomerators (water separators) should be located in positions where they receive sufficient engine heat to keep the wax in solution when the ambient temperature sinks below zero. This sort of thing can be carried too far, however, as engine performance can be seriously affected by excessive heat in the fuel system.

Because of the very nature of the diesel waxing problem it is essential that all parts of the fuel system which are likely to require attention should be as accessible as possible without the need to remove or disturb other components.

Manufacturers' recommended service intervals should always be adhered to as far as the *maximum* time or distance interval is concerned, but in the winter there is a definite case for increasing the frequency of servicing as far as the filters and water separators are concerned.

Methods aimed at preventing waxing-up of derv can be placed into

WELL DESIGNED SYSTEMS SHOULD INCORPORATE:

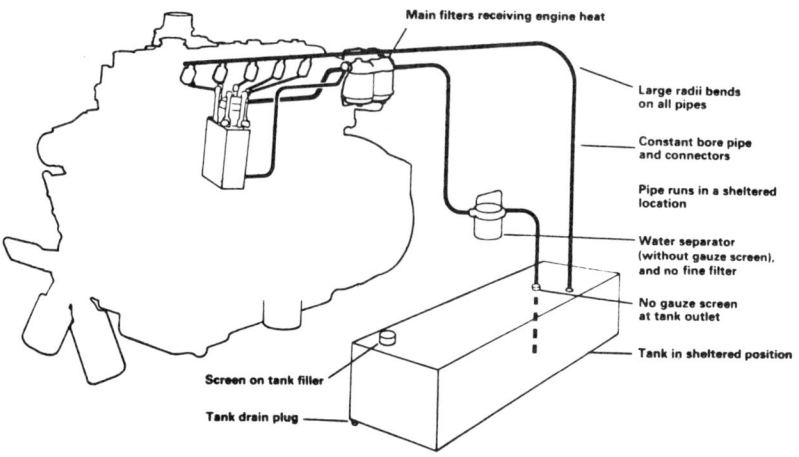

Figure 2.4 *The British Technical Council of the Motor and Petroleum Industries has published guidelines for fuel systems on board the chassis.*

BADLY DESIGNED SYSTEM

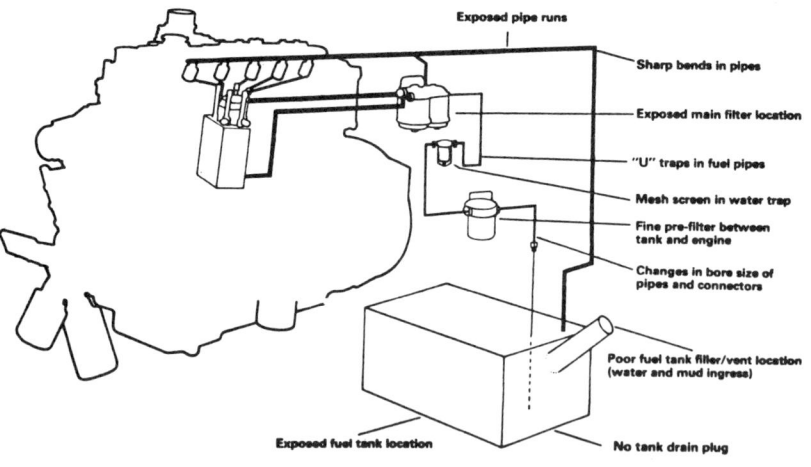

Figure 2.5 *A badly designed fuel system.*

HOW MUCH ARE YOU GETTING TO THE GALLON?

Lucas CAV offers protection throughout the fuel feed system.

By the time you come to use one gallon of diesel fuel it may contain enough contamination to completely wreck a fuel injection pump.

With the problems of dirt, water, and wax in diesel fuel you need protection right down the fuel line.

With over fifty years experience in designing, manufacturing and maintaining fuel injection systems, Lucas CAV offer a complete range of fuel conditioning products from the world famous CAV 296 filter – to the latest additions to the range, – the CAV Clip-On filter, Fuel Heater and Waterscan.

For further information contact:

Product Marketing
CAV Parts and Service Lucas CAV Limited
P.O. Box 36 Warple Way London W3 7SS
Telephone: 01-743 3111   Telex: 934986

Lucas CAV

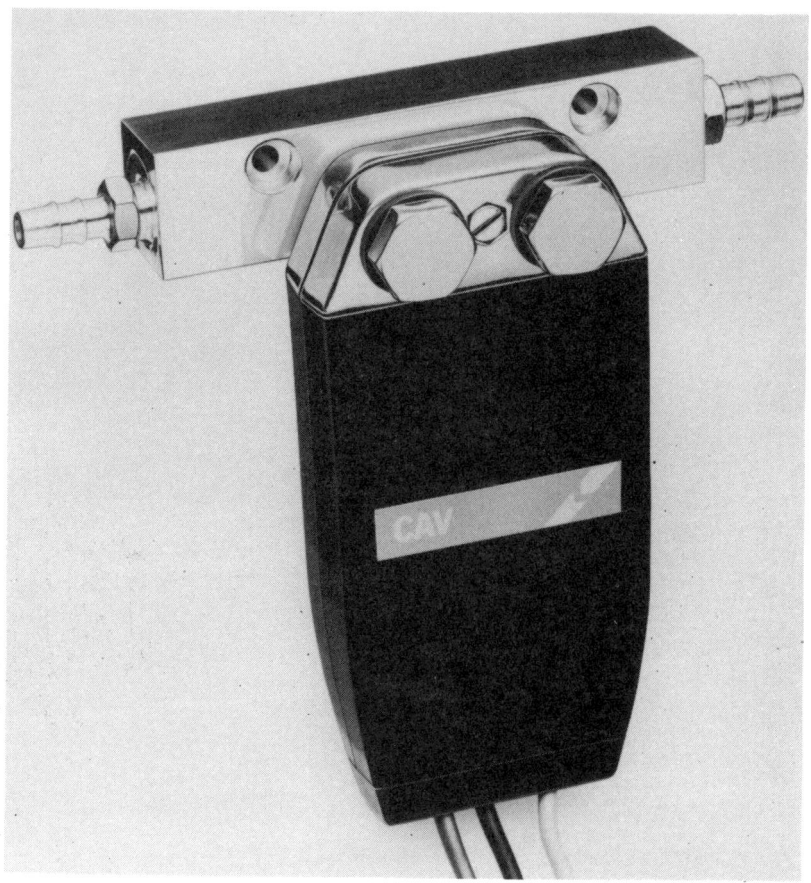

Figure 2.6 *The Lucas CAV fuel heater is claimed to give a minimum rise in fuel temperature of 10°C at a flow rate of around 10 gallons per hour.*

one or other of two categories: fuel heaters or fuel additives. Both are self-explanatory and both are well represented in the market place.

Lucas CAV Ltd offers the CAV Fuel Heater (see Figure 2.6) which is fitted in the fuel line and is suitable for flow rates of up to 2.3 litres/min (30gph). The output is 300W which CAV claims will give a minimum rise of 10°C at a flow rate of 0.75 litres/min (10gph). The output is controlled by thermostat and a warning light is provided to indicate when the heater is working.

The CAV Fuel Heater is priced at £49.22 with a typical fitting time being quoted as 1 hour.

Marketed by Antiwax Advancements, the AWA (pronounced 'Away') fuel heater takes the form of a wrap around heating element for the fuel filter. The printed circuit heating element is bonded and sealed to an aluminium base 260mm (10.2in) long by 50mm (1.97in) wide which is wrapped around the filter assembly and held in place by two heat resistant straps. Alternatively it can be bonded to a flat or curved surface, eg a fuel tank, using a suitable epoxy-resin adhesive.

Claimed to fit any size of filter, the AWA is available in a range of outputs from 10W to 80W. One of the main advantages claimed for this particular heater is that it can be used to melt wax crystals which have already formed in the filter, whereas most of the other heating devices are aimed at preventing the wax forming in the first place. In fact, the company claims that, for a cold weather breakdown recovery, the AWA can be fitted within minutes allowing the engine to be started after around 10 to 15 minutes warm up. The price depends upon the quantity ordered, starting at £25 for one AWA and falling to £16.50 for an order for over 100.

The Thermo Blend fuel heater from Fleetguard uses the relatively high temperature of the return fuel line to heat the cold fuel in the supply line. Intended mainly for the Cummins type of fuel injection system, the Thermo Blend will not melt wax which is already present as it relies on the engine running to provide its heat source. It is priced at around £50.

Raychem markets a fuel line heater known as Thermoline (see Figure 2.7) which replaces the existing fuel line and is available in 12V and 24V versions. It comprises a heavy duty hose with a self-regulating heating strip pre-assembled inside the entire length of the hose to protect the fuel line and provide warm fuel.

Raychem claims that if the driver preheats the fuel for about five minutes this will provide enough warm fuel to dissolve any residual wax in the filter. The Thermoline can be left switched on to ensure that no further waxing takes place in extremely low temperatures when the vehicle is moving. Depending upon the engine and its total fuel flow rate, the price of a Thermoline kit ranges from £116 to £149.

As mentioned earlier, the fuel companies do incorporate their own anti-waxing additives and this can cause difficulties in assessing the performance of the various proprietary brands of additive which, in part, account for the wide variation in claims made for each product.

The concentrations of additive required to give a reasonable degree of protection vary considerably from product to product, but on one point there is total agreement. To ensure satisfactory mixing the additive must be incorporated when the ambient temperature is still at least a few degrees above freezing. The best way of ensuring complete mixing is to put the additive into the bulk storage tank when the tank is nearly empty. The next delivery of derv will provide all the mixing that is necessary.

Figure 2.7 *The Raychem Thermoline comprises a heavy duty hose with a self-regulating heating strip incorporated throughout the entire length.*

The same principle applies on the road. If the vehicle is refuelled away from base then the additive content must be maintained. The additive manufacturers usually market their product in small quantities as well as large so it is not difficult to carry a one-litre can in the cab ready for topping up. Again this should be poured in just before refuelling when the agitation caused by the in-coming fuel and the subsequent vibration when on the move will be sufficient to ensure complete dispersal.

A list of the various companies offering anti-waxing additives is contained in the directory section of this book, Part 2.

The addition of kerosene to derv is a practical solution to the waxing problem but it can involve certain legal difficulties. Up to 25 per cent by volume of kerosene can give protection down to around -15°C (5°F) but the difficulty arises with HM Customs and Excise.

Because certain classes of fuel (eg for heating or agricultural purposes) are marketed as a duty rebated product it is not permissible for such fuel to be used for conventional road going commercial vehicles. To prevent rebated fuel being used for this purpose a dye is added which, while obviously not preventing such fuel being used, makes it very easy for such an offence to be detected.

Kerosene is a duty rebated fuel and as such it is a criminal offence in the eyes of HM Customs and Excise for it to be added to diesel fuel for vehicles intended for use on the public highway. However, in exceptional circumstances – and a paralysed fleet of vehicles might be included in this category – it is possible to obtain permission from the Customs and Excise people to add kerosene to diesel fuel. It should be stressed, however, that there is no guarantee that such permission will be given. It is only under exceptional circumstances that such approval is likely to be granted.

With duty paid kerosene (ie without the dye), its addition poses less of a legal problem and can be achieved in two ways. First, the fuel company can supply unmarked kerosene on which the full duty has been paid directly from bonded premises. Alternatively, the fuel company can blend unmarked kerosene with unmarked derv before delivery to the customer. The latter can only be done at bonded premises where the duty will be paid on the total blend.

The main point to remember when considering the addition of kerosene is that it will not dissolve wax already present in the fuel system. It is essential, therefore, that the derv/kerosene blend be fully mixed before going into the vehicle's fuel tank and even this does not ensure complete protection from waxing unless the associated pipework is drained and refilled with the mixture.

In short, the addition of kerosene is, in certain circumstances, a practical proposition although great care needs to be taken to ensure the legality of the operation.

Although some engine and vehicle manufacturers permit the addition of petrol to the diesel fuel in some circumstances, the practice cannot be recommended as it introduces a significant safety hazard. Petrol can give rise to concentrations of inflammable vapours in the vehicle's tank a long time after it was added, when everyone has forgotten it was ever used. In no circumstances whatsoever should petrol be added to the bulk derv storage tanks.

It cannot be emphasised too strongly that additives will not dissolve large deposits of wax already present in the system; they can only be of benefit by preventing the crystals from forming in the first place. It is a classic case of 'prevention being better than cure'. The same applies to the addition of kerosene or petrol to derv – with or without official approval.

If the worst happens and the vehicle is caught in a cold snap before any precautions have been taken, then there are a number of 'dodges' which may be tried. The fuel filter is one of the prime culprits and can often be unclogged by applying gentle heat. This means using something like a hair dryer or pouring warm water over the filter. A naked flame should never be used in an attempt to unwax a fuel system.

Caution should be taken to prevent hot water pouring over the injection pump. Different parts of the pump have different coefficients of expansion and warming up too rapidly can cause seizure because of the very fine tolerances used in assembly.

Finally, nearly the problems of this nature encountered during the winter can be avoided by careful routine maintenance and a little foresight.

3. The Design and Construction of Vehicles for the Carriage of Dangerous Substances

Brian Veale

For many years the only liquids carried by road and considered in any way dangerous were petroleum products having a flash point of below 21°C (73°F). Strange to relate, petroleum products with low flash points have been subject to stringent regulations since 1862, when an Act of Parliament was passed covering the safe-keeping of petroleum.

A hundred years ago an Act was passed concerning the hawking of petroleum and this can be considered the first regulation to cover the transport of petroleum products by road. At that time, the maximum quantity permitted was 20 gallons – barely sufficient to fill the tank of some of today's passenger cars.

Clearly, for over a century, successive governments have been aware of the potential hazards of petroleum spirit when being loaded, transported and discharged into customers' storage tanks. In that period we have seen permitted loads rise from 90 litres (20 gallons) to 36,000 litres (7,920 gallons), with maximum individual compartment sizes of 7,600 litres.

Today, it is estimated that in a typical year over two million deliveries are made in the petroleum spirit market at an average of 12,000 litres per delivery. The sheer volume of deliveries in Britain has meant that for many years the public has been aware of the potential hazards associated with petroleum and, in turn, regulations have been developed progressively to protect all concerned.

While all this was being formulated and put into effect, other far more hazardous products were being carried by road in many types of vessels built to a variety of specifications and often mounted on unsuitable vehicles.

For some 25 years the transport of dangerous goods in Europe has been covered by The European Agreement Concerning the International Carriage of Dangerous Goods by Road (ADR) which permits dangerous goods to be freely transported across frontiers provided that the packaging, labelling and means of transport comply with ADR regulations.

The tragic liquefied petroleum gas explosion which occurred in Spain some five years ago gave an added impetus to the formulation of more wide ranging and up-to-date regulations in Britain, and now many

Figure 3.1 *This lightweight ERF B-series tractive unit is coupled to a seven compartment 30,000 litre (6,600 gal) Thompson tank.*

substances are included in regulations which previously only covered petroleum liquids. To bring up to date the design, construction and maintenance of petroleum vehicles, road tankers and tank containers, in order to meet present day requirements, a code of practice is being drawn up by The Institute of Petroleum to augment the basic requirements of any road vehicle so that:

1. It is properly designed, of adequate strength and constructed of sound material.
2. It is suitable for the purpose for which it is being used.
3. The carrying tank is so designed, constructed and maintained to prevent the contents escaping.

The present day method of guiding and regulating the conveyance of hazardous loads by road is to draw up an approved Code of Practice to cover every aspect of the three areas of concern listed above.

An approved code of practice has a special legal position in that failure to observe it does not render anyone liable to civil or criminal proceedings, but allegations of failure to observe any part of an approved code can be admitted as evidence of a breach of law. Clearly the Code of Practice is a lengthy and comprehensive document, but it is interesting to highlight the following specific points which relate to vehicle chassis construction.

Vehicles for the Carriage of Dangerous Substances

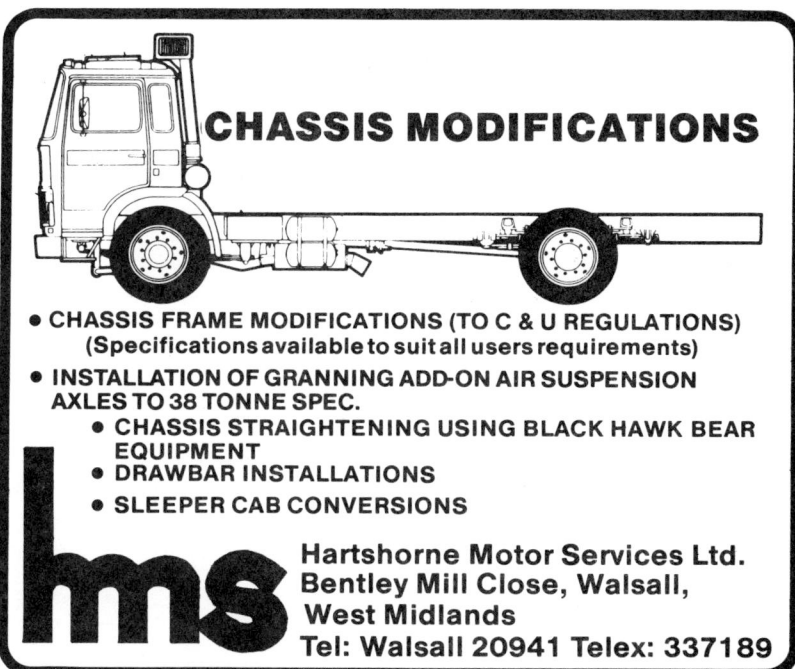

The Cab

The cab must be partitioned off from any dangerous substance to protect the driver and any passengers, ie a firescreen must be fitted and this firescreen must extend to the top of the chassis frame to prevent any product from spilling on to the exhaust system.

There should be no opening in the cab roof, and where windows are provided at the back of the cab they should not be capable of being opened. There should be a space of not less than 100mm (approximately 4in) between the rear of the cab and the foremost part of the carrying tank, or any attachment to the tank, other than an open-rung ladder and those connections between the cab and the tank which are necessary for the operation of the vehicle.

Engine and Fuel System

The vehicle's engine must be of a suitable type and installed so as not to cause risk to health or safety. Therefore an internal combustion engine is needed and, if gravity fuelled, a quick-acting cut-off valve should be fitted to the fuel line which should be clearly marked.

Figure 3.2 *The fire screen must extend to the top of the chassis frame and prevent any spillage of product on to the exhaust system.*

Vehicles for the Carriage of Dangerous Substances

REVERSING ALARMS

Minimise the risk of reversing accidents with the Triotronics reverse alarm. Its continual signal in wave form gives a distinctive sound designed to provide maximum warning to old and young alike – with widely varying hearing abilities. Accidents are expensive and time consuming – especially where personal injury is involved. Why wait until you've HAD an accident – It's cheaper to fit one now!

- **AUTOMATIC & MANUAL MODES**
- **TWO YEAR GUARANTEE** ● **BRITISH MADE**
- **SIMPLE TO FIT** ● **INEXPENSIVE**

Triotronics Ltd., 242a Eastwood Road, Rayleigh, Essex SS6 7LY
Telephone: (0268) 742089

Fuel Tank

This must be so placed as to allow leakages to drain directly on to the ground. If the fuel has a flash point of less than 65°C (149°F) then the fuel tank should not be behind the firescreen unless:

○ it is protected by guards or the chassis frame;
○ the filler cover is lockable;
○ the fuel feed pump is located in front of the firescreen.

Exhaust System

The exhaust system must be so designed and located as to prevent dangerous substances from coming into contact with it.

Air Intake System

The intake system must be designed, constructed and located so as to minimise the possibility of dangerous vapours being drawn into the engine.

Electrical System

Where there is a risk of fire and explosion the nominal voltage should not exceed 24V. Behind the rear of the driver's cab all wire conductors should conform to BS 6862:1971 (1981), to ensure that they are adequately insulated and are able to carry more than the designed circuit current without causing an unsafe rise in wire temperature. The conductors should also be adequately fixed and protected so as to minimise the risk of damage or deterioration.

Battery terminals should be effectively protected against the spillage of flammable liquid and should be insulated by a cover against inadvertent contact.

All articulated vehicles should have electrical continuity between the chassis of the tractive unit and the chassis of the trailer.

Where flammable liquids having a flash point of below 21°C (70°F) are conveyed, the following additional electrical requirements should be complied with:

Behind the rear of the driver's cab
1. No screw-in or capless bulbs should be used;
2. Junction boxes, connectors and all electrical equipment should be adequately protected and shielded as far as is practicable from the ingress of moisture or dangerous substances under normal conditions of use;
3. Electrical continuity should be maintained between tank and carrying vehicle; and,
4. Switches should conform to BS 4137: 1967.

Where an insulated return circuit is used, there should not be more than 300mA leakage between either polarity circuit and the vehicle chassis. A double-pole master switch to Zone 2 requirements, as defined in BS 5345: 1976, to enable all electrical circuits to be isolated (including open circuiting of the generator field windings), should be placed as near as possible to the battery. This should not prevent intrinsically safe circuits or flameproof circuits, to Zone 1 requirements, from being taken from the battery side of the master switch. The master switch control should be readily accessible to persons outside the vehicle and its location should be indicated by a clearly visible notice. Means should also be provided to enable the driver to open the switch without leaving his seat. A further visual indication must be provided to indicate when the master switch is in the 'ON' position.

Rear End Protection

The lower end of the carrying tank should be protected by a stout steel

11 litre turbo engine. 703 lbs. ft. of torque. Fuller 9 speed box. Almost seems a shame to stop for the night.

Leyland Trucks
Delivering the goods.

bumper which should be located at least 100mm to the rear of the tank shell.

Vehicle Stability

The height of the centre of gravity of the load should not be greater than 95 per cent of the distance between the outer walls of the tyres measured at the outside of their contact with the ground. In the case of an articulated vehicle the axles of the semi-trailer should not exceed 60 per cent of the total laden weight of the complete outfit.

Tanks

So much for the chassis, now what are the requirements for the tank?

It is the operator's responsibility to ensure that any tank being used for the conveyance by road of a dangerous substance, and any fittings attached to the tank:

1. Are designed, constructed and maintained so as to prevent any of the contents escaping, except that this requirement shall not prevent the fitting of a suitable safety device.
2. That tanks, in so far as they come into contact with the substance carried, are made of materials which are not liable to be adversely affected by the substance, nor liable in conjunction with it to form any other dangerous substance which could significantly increase the risk to the health and safety of any person. Tanks covered by the regulations fall into seven categories.

Type A tanks, being gravity discharged, are intended for the carriage of substances having, at 50°C (122°F), a total pressure (ie vapour pressure plus partial pressure of inert gases, if any) of not more than 1.1kg/cm^2 (15.7psi) absolute.

Type B tanks, which are pressure filled or pressure discharged, are intended for the carriage of substances having, at 50°C, a total pressure (ie vapour pressure plus partial pressure of inert gases, if any) of not more than 1.1kg/cm^2 absolute.

Type C tanks, whatever their filling or discharge system, are intended for the carriage of substances having, at 50°C, a total pressure (ie vapour pressure plus partial pressure of inert gases, if any) of not less than 1.1 and not more than 1.75 kg/cm^2 (25psi) absolute.

Type D tanks, whatever their filling or discharge system, are intended for the carriage of substances having, at 50°C, a total pressure (ie

WARN MODEL 8274 UPRIGHT 2-WAY REVERSIBLE 12V. & 24V. DC. ELECTRIC VEHICLE WINCHES.
Operation by remote control – automatic braking-free spooling clutch. Catering for single line load pulls of up to 8000 lbs

WARN MODEL 8274 2-WAY REVERSIBLE HYDRAULIC VEHICLE WINCH.
Automatic braking and free spooling clutch. Requiring hydraulic pressure of up to 1650 p.s.i. and flow rate of up to 12 g.p.m. for single line load pulls of up to 5000 lbs and winching speeds of up to 22 f.p.m.

MAXPULL MODEL GM5 & GM20 2-WAY REVERSIBLE HAND WINCH/HOISTS.
Both units have slow and fast winds, self actuated brake and safety latch. The GM5 Model catering for single line load pulls of up to 5500 lbs and Model GM20 single line load pulls of up to 10 tons.

WINCH DIVISION

Prices according to model. Details and quotations available on request:
215 KNOWSLEY ROAD, BOOTLE, LIVERPOOL L20 4NW
Telephone: (051) 933 4338

---

vapour pressure plus partial pressure of inert gases, if any) of more than 1.75 kg/cm^2 absolute.

Type E tanks, which must be of minimum thickness, and compartmented, are to be used for conveying petroleum spirit to premises licensed for keeping petroleum spirit for retail purposes only.

Type F tanks are to be used for road tankers and are constructed in reinforced plastics.

Type G tanks are to be used for road tankers intended for the conveyance of hazardous waste liquids.

Regulation 6 of the Dangerous Substances (Conveyance by Road in Road Tankers and Tank Containers) Regulations 1981 (SI 1981 No 1059) states that:

> 'the carrying tank of the road tanker or, as the case may be, the tank container and any fittings attached thereto:
> are designed, constructed and maintained so as to prevent any of the contents escaping, except that this requirement shall not prevent the fitting of a suitable safety device; and,

in so far as they are likely to come into contact with the substance, are made of materials which are neither liable to be adversely affected by the substance nor liable in conjunction with it to form any other dangerous substance which would significantly increase the risk to the health and safety of any person.'

Acceptable tank materials are: mild steel, stainless steel, aluminium (all to various British Standards) and reinforced plastics. In the case of reinforced plastics, minimum values are not prescribed for tensile strength but for the force which can be applied to the material. Tank shell thickness, baffle thickness and all constructional details are contained in the Road Tank Vehicle Design Code of Practice published by The Institute of Petroleum (address: 61 New Cavendish Street, London W1M 8AR). The details concerning the materials and constructional methods of all of these tanks are extensive and fully documented within regulations and codes of practice, and will be well known to all tank builders, who will clearly be able to give advice on all aspects of design and construction for the tanks listed above.

While regulations and codes of practice cover both chassis and tank construction requirements in great detail there are other aspects of chassis and tank equipment selection which are left to the person specifying the complete vehicle.

In these days of economic stringency there is a great temptation to keep the specification to that which just satisfies the legal requirement. However, safe operation over long periods with minimum downtime, ease of driver operation and ease of maintenance requires, where possible, a high specification.

Today's vehicles are expensive by any standards and very often the tank and equipment are more expensive than the chassis. Thus we must, in order to achieve economy, keep our vehicles for longer periods. Therefore whole life costs are very important, and low life costs, linked to safe operation, can best be achieved by careful specification.

Considering the chassis first, we must look for a specification which incorporates well tried concepts and units.

Engines today, particularly those made by the specialist engine manufacturers, are capable of giving long life, as much as 500,000 miles before a major overhaul. Power outputs are important and the engine must be well on top of its job. Therefore a power-to-weight ratio in excess of the minimum requirements should be specified. This should ensure both a long vehicle life and good fuel economy, particularly if the overall transmission gearing is high and the road speed limited, in top gear only, to say 62-65 mph.

The transmission selected can be according to customers' requirements, either a range-change or a splitter gearbox, or even semi or fully automatic.

Modern braking systems are much more safe and reliable, but here again a little forethought can help. S-cam brakes are most easily understood and serviced by garages and it is an easy matter to include in the specification some form of automatic slack adjustment.

At the same time we must not forget that the elimination of moisture in an air brake system considerably improves its safety and increases component life. Therefore, air drying equipment is a valuable extra. Suspension is another critical safety area and of the three types of suspension available, steel, air and rubber, it is fair to say that steel springs give the most trouble, air gives the softest ride and rubber gives the best all round ride and is generally fail-safe, ensuring that a failure of one component in the suspension system does not constitute a hazard and that the vehicle can still proceed carefully to its destination. With vehicles transporting dangerous substances this virtue is of prime importance.

A safe and effective electrical system is also important and in particular it is essential that we ensure rear lighting is of a good size and that lamps are fitted with dual bulbs on the important rear circuits.

Batteries are now more reliable than ever before, but topping up of batteries is still an unpopular chore, particularly if the battery box is located in an inaccessible place. Today, reliable low maintenance batteries are available and these should be chosen in preference to the older types, especially as neglect does not affect their performance or reliability.

Standardisation, therefore, must play an important part in the safety programme, particularly if vehicles are operated on a shift basis with several drivers using the same vehicle during any one period. Wherever possible one or at most two basic makes of vehicle should be operated, so that the cab layouts and controls are all located in the same place, which ensures that in an emergency the driver knows exactly where all the essential controls and switches are located. The time lost in trying to locate a switch in a strange cab at night can mean the difference between safe operation and an accident. We seem far away from the American idea of standard dashboard layouts but, wherever possible, operators should attempt to standardise on makes and layouts in order to increase the safety aspects of the vehicle. The additional benefits from standardisation on one model or make, are that maintenance staff can spot and correct problems more easily and that the stocking of spares is kept to a minimum.

It has been said that standardisation leads to stagnation – perhaps, therefore, the word 'commonisation' would be more appropriate. Regardless of the size and capacity of vehicles used for the carriage of dangerous substances, commonisation can play a large part in increasing safe operation without incurring increased costs. Typical examples of areas which can be similar on tank vehicles of all sizes are:

- ○ access ladders
- ○ walkways
- ○ brake and electrical connections
- ○ rear end protection
- ○ fire extinguishers
- ○ labelling of controls

Access Ladders

These are important safety features and the choice of front or rear mounted ladders can be left to the user, but the rear mounted ladder is safer laid out as it can be taken nearer to the ground. It is important that the ladder is not vertical but is at a slope of at least ten degrees from the vertical. The rungs should be evenly spaced and as wide as possible. At the top of the ladder a safety platform canassist drivers and loaders in ascending or descending (the Health and Safety Executive have published guidance notes on access to road tankers (Guidance Note GS26).)

Walkways

Vehicles used for petroleum spirit conveyance have for a number of years had to have roll-over protection coamings to safeguard the lids and tank top fittings. Consequently, it is an easy matter to cover these areas with wide walkways made of a non-slip material, but it is important to ensure that all tank top protrusions are kept below walkway level wherever possible to prevent drivers from tripping and falling from them. Whatever the type of tank, walkways should be as wide as possible. An additional advantage can be gained if dipsticks are carried horizontally and recessed below walkway level as, apart from reducing the risk of drivers tripping on them, breakages can be reduced.

Brake and Electrical Connections

Brake and electrical connections between tractors and trailers are important safety areas and care must be taken to ensure that, whatever type of trailer is connected to the tractor, the brake and electric power lines are located in such a position that they do not foul any part of the tractor or trailer.

Apart from the clear labelling of brake pipes it is a good idea to label the electrical connection layout clearly so that should the plug be pulled from the wiring it can be correctly rewired without delay. Brake pipe colour coding should be consistent throughout the vehicle; the present day use of plastic piping helps considerably *provided* that it is not completely painted over, as is often the case.

Rear End Protection

Rear end protection is now mandatory and has now been augmented by the need to provide additional under run protection. Mounting the main bumper as low as possible reduces the risk of rear end damage in most accidents, and it also provides a useful intermediate stage in the access ladder layout.

Fire Extinguishers

The provision of fire extinguishers is also compulsory and, if possible, it is advisable on large vehicles to provide more than the bare minimum number and sizes of extinguishers as stipulated in regulations.

Extinguishers are vulnerable to weather and road conditions and take a considerable pounding on tankers, which invariably travel unladen for at least half of their working life. They should, therefore, be located in secure, weatherproof containers suitably identified and labelled and located away from the main hazard areas, where spillage or fire might occur.

In passing, it should be stressed that extinguishers of the powder type are not 'fit and forget' items. They must be checked and serviced regularly in accordance with makers' recommendations.

Labelling of Controls

Spillages and incorrect deliveries are still caused by poor and inadequate instructions to drivers on the correct use of a vehicle's delivery system.

Clear, concise labelling of the basic system is a major contribution to safe operation. Where the loading and discharge system is more complex than the typical multi-compartment tanker, the provision of a flow diagram mounted near the controls is essential to reduce the risk to user and customer.

Conclusion

These are some examples of how safety can be greatly increased at little or no extra cost and these can be adapted and extended by the user with careful thought and by asking the views of drivers. Much valuable information can be given by drivers in respect to controls layout and delivery pipes, etc. Safety can be included in the design, specification and construction of every vehicle, but until it is loaded and used on the road no risks are involved. We must not lose sight of the fact that throughout its working life a vehicle needs further safety measures built into it at regular intervals through correct and effective maintenance.

Also, while users can take advantage of the chassis makers' servicing schedules, the maintenance of the tank and ancillary fittings must not be neglected. Where these are concerned the user must prepare his own schedule of maintenance, based on experience, and ensure that this is rigidly adhered to and incorporated in the overall vehicle servicing schedule.

The safe operation of road tank vehicles is essential to all sectors of the community and while the regulations and codes of practice lay down the basic requirements for the construction of vehicles used to convey dangerous goods, many additional safety devices can be incorporated into vehicles by simply using a little extra foresight at the specification stage, and then continuing to use and maintain vehicles with safety always in mind.

4. Vehicle Noise Legislation

Graham Montgomerie

In December 1980, the 'Report of the Inquiry into Lorries, People and the Environment' was published. This made public the findings of the committee headed by Sir Arthur Armitage which formed the basis for the 1983 legislation on higher gross weights. Not all of the Armitage Report was concerned with heavier weights however – the environmental issues were given just as much priority and one such issue was that of vehicle noise. The relevant paragraph stated that:

> 'The Government should adopt as explicit aims of policy that an EEC Directive should be agreed requiring lorries to be manufactured to a maximum noise level of 80dB(A) and that the new level should be introduced as soon as practicable in the light of discussions with manufacturers and operators, and not later than 1990.'

It is this extremely difficult target which is causing concern to all vehicle manufacturers as to how they can achieve this figure without an excessive penalty of cost and weight.

The Decibel

Vehicle noise legislation revolves around the unit of measurement known as the decibel which is a relative measurement or ratio rather than an absolute value. The Bel was originated in the 1920s to describe the attenuation of signals in telephone cables, and one Bel is the ratio R given by:

$$\log_{10} R = 1$$

Because this is a large ratio, a smaller one, the decibel, is more widely used, where one decibel is the ratio r given by:

$$\log_{10} r = 0.1$$

What these logarithms mean in practical noise measurement terms is that an increase of 10dB is experienced by the average listener as a doubling of the loudness (see the table below).

The decibel is a measure of the sound pressure level and ranks noises

Noise stage	Type of noise	Loudness	Perception
30-65dB(A)	Whispering, rustling of leaves	30dB(A)	Very quiet
	Quiet residential street	40dB(A)	Fairly quiet
	Normal domestic conversation	50dB(A)	Normal
	Office noise	60dB(A)	Normal
65-90dB(A)	Loud conversation, shouting, car at 10m distance	70dB(A)	Loud
	Road noise with heavy traffic	80dB(A)	Loud
90-120dB(A)	Noisy factory shop	90dB(A)	Very loud
	Car horn at 7m distance	100dB(A)	Very loud to unbearable
	Aircraft engine	120dB(A)	Very loud to unbearable

Note: *The threshold of pain is reached at 130dB(A)*

in terms of pressure, but taking no account of the ear's decreasing response at low and high frequencies. In an attempt to duplicate more accurately the response of the human ear, sound level measuring instruments are often fitted with frequency weighting filters known as A, B and C. Experience has shown the A weighted decibel to be the most suitable unit for measuring vehicle noise and this is usually expressed as dB(A).

Although the report of the Wilson Committee in the early 1960s made some recommendations aimed at reducing traffic noise it was not until 1970 that noise limits for all new vehicles were introduced for the first time. The limit was set at 89dB(A) for both trucks and passenger vehicles and was measured in accordance with the test procedures laid down in BS 3425: 1966.

Testing Procedure

Within this standard, a number of environmental and dimensional conditions are laid down before any practical testing is attempted. For example, the level of ambient noise (including wind noise) must be at least 10dB below that likely to be produced by the test vehicle. In this context, gusts of wind can often disrupt a test when other conditions are thought to be satisfactory.

As the presence of people can influence the meter reading to quite an appreciable extent, it is usually suggested that a maximum of two people should be allowed to remain in the vicinity of the vehicle or the microphone.

Vehicle Noise Legislation

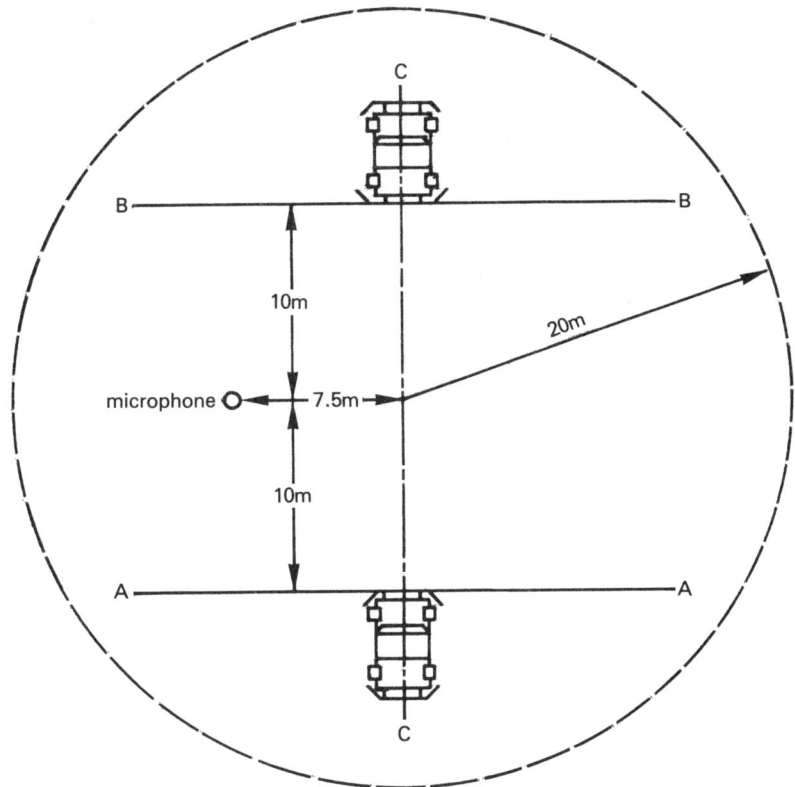

Figure 4.1 *The dimensional requirements for the noise test area are shown here. The open space containing this area should have a radius of at least 50m to minimise the risk of interference from buildings and trees.*

The dimensional requirements of the test area are explained in the diagram below (Figure 4.1). Although the central test area is contained within a circle having a radius of 20m (66ft), the open space containing this test area should have a radius of at least 50m (164ft) to minimise the risk of having the results distorted by the presence of walls, buildings and even trees.

In practice most manufacturers give themselves a wide margin of safety and treat this 50m dimension as very much the minimum figure.

The approach speed procedure for BS 3425: 1966 is somewhat complicated and needs some explanation in the light of the new test procedure being recommended.

If the vehicle has less than five forward gears then second gear should be selected, whereas third is used for a gearbox with more than five gears. If the specification includes a two-speed axle or a splitter, then the 'high' ratio should be used.

The actual approach velocity should be at an engine speed of three-quarters of the speed at which the engine develops its maximum power, *or* three-quarters of the engine speed permitted by the governor, *or* 50km/h (31mph) – whichever is the lowest.

The vehicle is driven up to line A-A (see Figure 4.1) at this previously determined speed and in the appropriate gear. When the front of the vehicle reaches the line, the driver accelerates at full throttle across the test area until the rear of the vehicle crosses the finish line (B-B) of the test area when the throttle is then released.

If the driveline characteristics are such that the governor comes into operation while the vehicle is still passing through the test area then the next highest ratio is used to prevent this occuring in a repeat test. Three noise readings are taken in each direction to give six in all.

Quiet Heavy Vehicle Project

One of the major steps taken in an effort to reduce vehicle noise came with the Quiet Heavy Vehicle project which was a joint exercise carried out by the Transport and Road Research Laboratory, the Motor Industry Research Association and the Institute of Sound and Vibration Research at Southampton University. From the manufacturing side, Leyland, Foden and Rolls-Royce were also involved.

The principal objective of the QHV project was to produce two vehicles; one was to be for operation at 32 tons gross with a power output of 158kW (212bhp), while the other was to be capable of handling 44 tons with a power output of 261kW (350bhp).

The target noise level was set at 80dB(A) and this was to apply not only under the conditions of BS 3425: 1966 but also under 'any normal operating conditions'.

The 32 ton vehicle was intended to be a Leyland Buffalo powered by the turbocharged 510 engine, but Leyland withdrew at an early stage because of the company's long-term plans to discontinue production of this power unit. Thus the QHV project entailed developing the 44 ton version only.

The first Foden test chassis was powered by a 261kW (350bhp) version of the Rolls-Royce Eagle engine. At that time this particular Foden chassis was about to be phased out but, rather than wait for the later model, work on the exhaust and cooling systems was begun on the older version with the technology being incorporated into the later 4 × 2 Foden which was plated for operation at 38 tonnes.

The engine for the later Foden was the 240kW (320bhp) version of the same basic Eagle design but it was modified considerably in certain major areas. The crankcase was divided horizontally through the crankshaft axis with the main bearing caps incorporated in the lower portion to increase the structural stiffness of the whole unit.

The timing gear train was re-sited at the rear of the engine, as the torsional activity is less at the flywheel end. The actual gears themselves were changed to a helical design in contrast to the normal Rolls-Royce practice of using straight-cut gears.

Sound damping cladding was bolted to the external surfaces of the engine including the block.

The cooling system was the most difficult part of the exercise and incorporated a fan which was designed by the National Engineering Laboratory at East Kilbride in conjunction with Marston Radiators. It was of the mixed flow type which combined mixed flow with axial flow and was driven hydraulically by a pump mounted on the engine. Because the radiator and fan took up a lot of room (the former was in fact 16 per cent bigger than the then current production Foden), the engine was moved back in the chassis by several inches.

The silencer was mounted transversely in Petroleum Regulations fashion and consisted of two 1.8m (6ft) long cylinders with a diameter of 245mm (10in). No absorption material was used – damping being achieved using tuned pipe lengths.

The cab itself incorporated masses of insulating material with sound absorbent matting being used for the bulkhead and the floor and insulating foam being pumped into the roof panels.

When tested to BS 3425: 1966 the Foden QHV (see Figure 4.2) recorded 80dB(A) which, to put this result into perspective, is about the same level as a petrol-engined private car. This was achieved with a weight penalty of 250kg (5cwt) and a cost penalty which was estimated to be around 10 per cent, making the basic assumption that the modifications were costed as standard production items and not experimental 'one-offs'.

The design of this vehicle has been covered in some detail as its performance in reaching 80dB(A) has influenced government thinking on vehicle noise legislation ever since.

At the same time as the QHV project was being carried out, it was proposed that new vehicles powered by engines with outputs in excess of 200bhp should be limited to 84dB(A) by 1 April 1974. Although UK vehicle and engine manufacturers invested a considerable amount of time and money towards this target, the entry of Great Britain into the European Community put a stop to the proposed legislation as it was not possible to have national legislation more stringent than that applicable in the EEC. The directive in force at that time (70/157/EEC)

Figure 4.2 *When tested to BS 3425: 1966, the Foden QHV recorded 80dB(A) which is about the same level as a petrol-engined passenger car.*

limited trucks and buses to 89dB(A), with a higher limit of 91dB(A) for vehicles having engines capable of over 200bhp.

Although the UK (along with France) continued to press for stricter noise control it was not until 1977 that a new directive (77/212/EEC) was completed to come into force for new vehicles manufactured from 1 April 1983 and first used after 1 October 1983. For heavy lorries, above the 200 horsepower level, the limit is now 88dB(A) (the full list of noise limits is shown in the table below). Note that the noise legislation is far more strict relating to passenger vehicles.

The latest development in European noise legislation has been the amending Directive 81/334/EEC which, although it has the same noise limits as in 77/212/EEC, introduces what is hoped will be a more realistic test procedure as the legislators felt that the old method did not reflect the true noise situation in an urban environment.

With this latest test procedure, the basic acceleration through the noise zone is the same as the dimensions of the zone, but the difference is that the vehicle is tested in every gear in the top 'half' of the gearbox. In other words, a vehicle with a 12 speed gearbox will be tested in each gear from (and including) seventh and upwards, with the noisiest case taken as the noise level for that particular vehicle. The approach speed of the vehicle as it nears the box remains the same.

Vehicle Noise Legislation

Development of EEC Noise Legislation

Vehicle category		EEC maximum noise limit values (dB(A))			
		EEC 70/157	EEC 77/212	EEC 81/334	EEC long-term proposal COM (83) 392
N1	Goods vehicles not exceeding 3.5 tonnes gvw	84	81	As for EEC 77/212 but with a change in test procedure	78
N2	Goods vehicles exceeding 3.5 tonnes gvw	89	86		Engine power — Up to 75kW: 80; 75 to 149kW inclusive: 82; 150kW and over: 84
N3	Goods vehicles exceeding 12 tonnes gvw and engine power of 200bhp (DIN) or over	91	88		New 75kW threshold introduced, 147kW threshold changed to 150kW, 12 tonne threshold deleted
M1	Passenger vehicles up to nine seats including driver	82	80		77
M2	Passenger vehicles over nine seats including driver and not exceeding 3.5 tonnes gvw	84	81		78
M2/M3	Passenger vehicles over nine seats including driver and exceeding 3.5 tonnes gvw	89	82		80
M3	Passenger vehicle over nine seats including driver and engine power of 200bhp (DIN) or over	91	85 — 200bhp threshold changed to 147kW		83 — 147kW threshold changed to 150kW

EEC Noise Proposals COM (83) 392

Vehicle type	Power output	Maximum noise level
1. Passenger cars	—	77dB(A)
2. Buses and coaches	150kW 150kW	80dB(A) 83dB(A)
3. Minibuses and vans (3.5 tonnes grw)	—	78dB(A)
4. Trucks	75kW 75-150kW 150kW	80dB(A) 82dB(A) 84dB(A)

Note: The engine power is quoted throughout in EEC legislation in kW but, as a guide, 150kW is equivalent to 201bhp

Initial testing would suggest that this change in test procedure can result in a noise level up to 3dB(A) worse than that achieved to BS 3425: 1966. Thus, the QHV would be unlikely to achieve its much quoted 80dB(A) under the new testing method.

Although not yet confirmed, it is likely that 81/334 will come into effect for vehicles manufactured from October 1985 onwards and registered after April 1986.

For the long term, the EEC is proposing further noise reductions which have been issued under COM (83) 392. Because this makes a number of radical changes to the noise limits in the different categories, the document must get the approval of the European Parliament before it is published in Directive form. This is expected in mid-1984 for implementation around 1988.

The full list of the proposals put forward in COM (83) 392 is shown in the table above with the major sources of worry to the manufacturers being the 83dB(A) for buses and coaches over 150kW (201bhp) and 84dB(A) for trucks with engines exceeding this power figure. These noise limits, of course, will be related to the new testing procedure.

Noise Sources

Whereas the manufacturers tend to agree on which areas of the vehicle make the biggest contribution to the noise level, there is less agreement on how a reduction should be achieved. The analysis of drive-by noise has shown the drive train in total to be the worst offender with the engine in particular needing the most attention for further improvement.

In general, engine manufacturers have opted for turbocharging which offers the most benefits on balance towards meeting legislation, although this, of course, was not the primary aim of adopting forced induction. Improved fuel consumption, increased power, reduced emissions and reduced thermal loadings are all benefits claimed for incorporating a turbocharger. As far as noise legislation is concerned, the turbocharger makes it possible to retard the timing and thus achieve a 'softer' combustion without a power penalty.

The proposed noise regulations are likely to cause more trouble for the passenger vehicle industry than for freight carrying vehicles and, with the former, experience with turbocharging has not been completely satisfactory. Because of the inherent lag of a turbocharger caused by the inertia of its rotating parts, the unit has not responded well to transient conditions which made it less than ideal for stop/start operation – a particularly serious defect for stage carriage bus operations. A lot of the early difficulties could be put down to using what amounted to a truck turbocharger for bus operation, when the whole pattern of utilisation was completely different.

Another problem associated with the turbocharger is that of its installation. It is an extra component which requires space and, what is more, it is a hot component. With a truck the temperature of the turbocharger does not cause many installation difficulties but with a passenger vehicle the engine tends to be located in a fairly confined space anyway and this situation will worsen if encapsulation is required.

Turbocharged engines do demand higher standards of maintenance than naturally aspirated units in the areas of air filtration and lubricant supply, to the extent that the National Bus Company has been quoted as spending over £500,000 per year on maintaining turbochargers alone on a fleet of 4,000 vehicles.

Although it can be seen that the turbocharger is not without its problems its use will increase with trucks and, in particular, with buses and coaches if noise and exhaust emission legislation is to be satisfied.

Encapsulation is another method of reducing noise levels but, at the moment, this is regarded by vehicle and engine manufacturers alike as being very much a last resort. The principle of full encapsulation is just what it sounds like: the engine is enclosed in a sound damped box which brings with it all sorts of attendant cooling and maintenance problems. Because of the temperature reached by a running engine some form of forced ventilation is required and great care is needed to ensure that this ventilation does not itself become a primary source of noise.

The panels of the encapsulation would need to be double skinned, incorporating some form of noise absorbent material. This particular area is giving concern to both manufacturers and operators because of

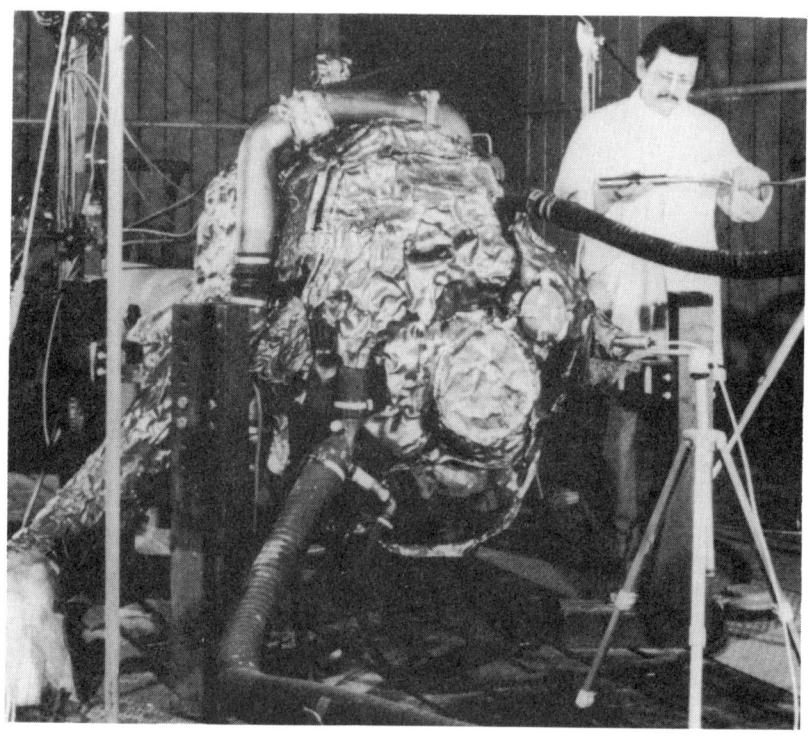

Figure 4.3 *Encasing the engine in lead, except for certain specific areas, is a useful way of isolating the major noise sources. Underneath all this is a Renault engine!*

its potential as a fire hazard. Most of the absorbent materials currently under development are designed not to absorb oil because they have a thin skin surrounding the body of the insulation. When new, the materials repel oil but it is far too easy to damage such insulation in service with the result that oil or derv could be absorbed.

The main objection raised against total encapsulation is that it is impossible to carry out a cursory inspection. With the maintenance systems operated by many fleets, a quick overnight inspection in the form of a 'cursory glance' is often employed and this would be impossible with full encapsulation.

By its very definition, the success of total encapsulation depends entirely upon the word 'total'. If some of the panels are not replaced – after removing them for routine access perhaps – then the whole benefit is lost.

Changing to total encapsulation will certainly add to the weight of a vehicle. Taking a bus as an example, Leyland has claimed that to reduce

Vehicle Noise Legislation

Figure 4.4 Anechoic chambers for noise testing are no longer a luxury; ever-tightening European legislation on vehicle noise has seen to that.

the noise level from 89 to 82dB(A) takes approximately 200kg (4cwt) of insulating materials. To reduce the noise by the next 3dB(A) will require the addition of a further 200kg.

Encapsulation is likely to add some £3,500 to the initial cost of the vehicle while its effect on maintenance costs has been put at anything between £200 and £400 per year per vehicle.

If all the proposed noise legislation does become a legal requirement then the vehicle manufacturers will need more help from the component suppliers, in particular the power unit manufacturers. More analytical work will be required on the basic design of the engine scantling itself to reduce the noise at source rather than merely contain it by encapsulation.

Noise is not safety related and thus cannot be compared with, say, the requirement to fit side guards. Thus, beyond a sensible level, every decibel becomes very costly. This is undesirable in many ways, but at least it is a situation applicable to all European vehicle manufacturers. The worry here relates to the export markets where there are no regula-

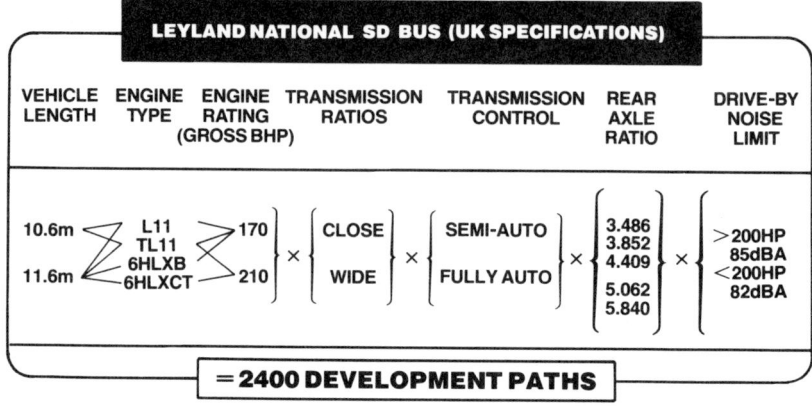

Figure 4.5 *To achieve the required noise limits it may be necessary to restrict the number of vehicle options. The Leyland National is available in 2,400 'different' specifications.*

tions governing noise and where the financial status of the countries involved limits their ability to pay extra for quiet vehicles.

If more cladding or encapsulation is required then it must be as durable and cost effective as possible as noise abatement cannot be sold to the operator as an environmental advantage and this is going to become a major marketing problem for manufacturers. To achieve the required noise limits it may be necessary to restrict the number of variants available. Consider the case of the Leyland National integral bus illustrated in Figure 4.5. The choices available (which are not untypical) give a total of 2,400 different specifications.

Certain combinations of engines and transmissions are likely to make it difficult, or even impossible, to meet the noise legislation. Using a passenger vehicle specification to illustrate the point, a bus with a naturally aspirated engine, a semi-automatic gearbox and a low axle ratio would be the most difficult to quieten compared with the quietest specification of a turbocharged engine, a fully automatic gearbox and a high axle ratio.

Some types of passenger vehicles present more difficulties than others, not so much because of the detailed specification of the components but because of how those components are installed. A rear-engined chassis with a horizontally mounted engine is a good example. Because of the projected area of engine to the road, the reflected noise is far greater than with a vertical engine. In this case the type of road surface becomes even more critical as asphalt, for example, is about 1dB(A) quieter than concrete. If such buses could be tested on soft snow a few of the problems would disappear.

5. Light Van Selection

John Parsons, Technical Editor, *Motor Transport*

Light van selection is, in theory, a quite simple process, but, in practice, a complicated one. Most users have some idea of the general size of vehicle they want, although it tends to be called a 'Transit type' rather than defined by any need for a particular gross weight or cubic capacity. That technique has served the user tolerably well for a long time but if used nowadays, with such a bewildering array of different vans and pickups on the market, it is too easy to end up with the wrong choice. For every weight sector there is now a choice of front or rear wheel drive, standard or high volume carrying capacity, and petrol or diesel engined options. It is now possible to buy a van which is sturdy, comfortable, economical, fast and reliable as well as popular with its drivers. Such a combination of factors was unheard of until recently. But it is still possible to make a big mistake in the selection of the right specification.

So where do you start? So few users have the slightest idea of what they really require that a few minutes spent in the traffic office can prove very valuable. The van that was purchased a year or two ago for those big parcels contracts is probably now carrying letters and documents. The single Transit van bought for the service bay cannot now carry all the special tools as well as the service crew so two vehicles are now required. And that cheap pickup you acquired last year is now so uncomfortable to drive that everyone from the works manager down is trying to wear it out as soon as they can.

If any of this sounds at all familiar, then you are going to have to sit down and think out your specification. You will waste money for every day you have an unsuitable van in the fleet. Buying a $10m^3$ ($13yd^3$) capacity van for a need which might arise only twice a year makes no sense at all, except perhaps to the manufacturer. And yet it is a common method of beginning a list of specification requirements.

To do this properly a list should be drawn up to include all the vans that fall into your requirement range – the list should include those vehicles which you do not like as well as those you do. Only in this way can it be proved that the final choice is a sensible one. But be warned, you could be in for a few surprises.

The basic Ford Transit is slightly longer than the basic Bedford CF

Figure 5.1 *The modern trend to front wheel drive in van design can give a low flat floor along with a useful height and volume for a given length. This is the high top Fiat Ducato.*

van, although the CF can handle a larger payload and has greater cubic capacity than the market leader, the Transit – but the Transit 80 is nearly £200 cheaper than the CF 230. The temptation to take the cheapest is strong but, if you need regularly to transport one tonne loads, there may be occasions when you will have to use two vehicles to stay within the law, in which case the £200 saving disappears. So buy a bigger Transit, the standard 100, for example. As a result, you can obtain the extra payload, while retaining the same cubic capacity for an extra £400. At the other end of the scale the Fiat Fiorino van packs in nearly twice as much cubic capacity as the similarly powered Volkswagen Golf van and the two are of almost identical length. In this case the Fiat is about £700 cheaper. Yet the Golf sells well both here and in France, the two markets where vans are cheaper because of tax conditions. Strange things happen when you start looking deep into specifications and price tables.

You can play a similar game with cubic capacity as a proportion of overall length. This is a measure of how well the manufacturer has used the road space that his product occupies. Why should you, the user, have to pay for the vagaries of a design that can result in more space being allocated to the engine and transmission than to the cargo? In many car-derived vans the average payload space is no better than 40 per cent of the overall length. In this category the latest range of front wheel drive vans stands out as giving very good value for the space they provide. The Fiat Ducato and the Talbot Express – both with the same basic design – have a deck length of 2.66m (8ft 9in) in an overall length

Light Van Selection

Figure 5.2 *The Renault range of vans is alone in being available in front or rear wheel drive versions. This standard 3.5 tonne Master shows what a low rear loading height is possible by keeping the drive up front.*

of 4.76m (15ft 7in). Compare this with any of the more conventional forward or semi-forward control panel vans and the proportion is slightly worse. The lower floor line of this layout also enables the Fiat/Talbot design to give a cubic capacity of $6.5m^3$ ($8.5 yd^3$) in this length, compared with the usual $5.5m^3$ or so in other designs. The same is true of the Renault Trafic and Master range that also feature rear loading heights of less than half a metre in their front wheel drive versions. Yet the overall height of most standard roof vans is about the same at around 2.0m (6ft 7in). Since all manufacturers choose their designs to minimise cost and maximise profit, you need to make sure that you do the same. If you do not require a high cubic capacity van then do not buy one. Salesmen are very skilled in highlighting those

79

Figure 5.3 *The car derived van is a strictly British phenomenon because of the tax position. Cubic capacity is strictly limited in most designs such as this Ford Fiesta. Most of the road space is not cargo carrying, but performance is usually excellent.*

details of the total specification that best flatter their own product. So make sure you have all the facts available to you. Several publications have lists of van types available which quote most of the figures necessary to beat the salesman at his own game.

One particular area to watch out for is the quoting of figures for payload capacity. Traditionally, this is calculated by taking the permissible gross weight and subtracting the unladen weight of the van. This is all very well if you are going to carry a single high density package – rather a rare event. Often, you will find that the maximum payload quoted assumes that the load will be placed so as not to overload an axle, and that, therefore, rather greater amounts can be carried than if the load is uniformly distributed. The Astra van, for example, has an excellent gross carrying capacity of 525kg (10.3cwt), but has been revised recently to raise the uniform load rating to 415kg (8.2cwt) by increasing the rear axle load. Asking which load rating is being quoted is one way of making sure that you know what you are getting.

Can you specify the degree of comfort? This factor has in the past not loomed large in operators' minds, until it began to dawn on users that a comfortable driver is a safer one, and that these days we all have a right to a tolerably comfortable workplace. The frustration that can occur with an awkwardly placed switch can be quite sufficient to cause the driver to break it off. A driver may also misuse a van he does not like. This is typical of human nature and is a factor which must be taken into account. No amount of official instruction to the contrary is going to make any difference to a driver who has spent all day with an aching back caused by a seat which will not adjust properly. This means that

Light Van Selection

Figure 5.4 *A low loading height is not necessarily the only consideration. The available opening width is just as important. This is the Talbot Express which has done particularly well in Britain, and is the fruit of a cooperative venture between three manufacturers.*

both you and your driver should try the vehicle – not an unreasonable demand. But do not be misled by imagining that 'all Transits are the same'. They certainly are not.

A 1984 Transit or Sherpa van is not the same vehicle it was when it was first introduced. In the Sherpa's case it is much more comfortable and a lot quieter too. The Bedford CF has become easier to handle and quieter but now shows its age in comparison with some of the more recent models. The layout of the drive, whether at the front or rear, is of far less importance for comfort than how the remaining area is designed around the power unit.

Good points, where comfort is concerned, are independent front suspension, a high integrity structure that does not rattle, boom or shudder and a level of noise intrusion that makes motorway driving tolerable. All these points conflict in some way or other with the need of operators for the least first cost. He will try and insist on high reliability of components and will endeavour to do without anything that does not actually earn money. Times have changed as far as van design is concerned but not all operators have changed with it.

Take front wheel drive for example. The advantages are nearly all in

Figure 5.5 *Bedford was one of the first manufacturers to add luxury to working vans, by including carpets as well as rear wash/wipe and heated window on the Astra. This also went faster than any van to date, with 105mph available from the 1600 version.*

favour of the daily user and nearly all against the workshop staff who have to maintain it. The low floor height and the compact drive arrangement, which leaves more room for the feet while keeping all mechanical components in one place (which helps with noise reduction), are all popular with today's van user. But the spectres of high service costs, extra complication and less out-and-out ruggedness still exist. True, the accessibility is often a challenge to the average fitter with only normal workshop tools, and a good old fashioned leaf spring suspension is very cheap to maintain as well as being strong, but do these points matter if the basic reliability is there? For that is precisely what the driver, fitter and buyer are looking for. A regrettable fact of life is that once a particular mechanical solution has got a bad name, it will keep it until a new generation of fleet engineers with different experiences comes along.

Meanwhile, it remains true that the high comfort, modern, front wheel drive van design is slowly making inroads into the traditional front engine, rear wheel drive, leaf sprung arrangement. The latter is still the preferred layout for the larger capacity urban delivery vehicle which operates at, or close to, its maximum design weight for long periods. But you have only to look in the high street or small shop

Light Van Selection

Figure 5.6 *The Volkswagen LT has a useful square shape and has kept well up to date over the years. The range now has six-cylinder engines throughout, which aids performance and adds smoothness.*

delivery bay to see how well the other 'system' is doing. The Fiat, Talbot and Renault ranges have stepped in where there was only the Transit or Sherpa previously. The revisions announced recently to the Transit, Sherpa and the Bedford CF ranges all serve to show that the big British-based makers are having to work hard to keep their sales up – and the Transit replacement is not due until 1986, when the old concept will be 21 years old. One wonders whether Ford has yet made up its corporate mind to go for front wheel drive, as there are still many conflicting opinions on this question at Ford. Renault solved this by offering a choice of front or rear wheel drive on the Trafic and Master. So, provided that reliability and maintainability can be improved still further, there seems no reason why the front wheel drive system should not take over. That is the way it began with cars, and the argument still rages about which system is best, although it would be hard to think of the Astra van ever having been thought of as anything but front wheel drive. The lead it gave in comfort, handling and general driveability as well as reliability in its class makes it strange to recall how, in the days of the old Escort, the Bedford HA and the Marina, few product planners thought that front wheel drive would ever sell. There was the Minivan of course, but that probably only served to emphasise some of the disadvantages.

The fact that the Renault Master is still the only volume vehicle manufacturer with a front wheel drive option at 3,500kg shows that the

Figure 5.7 *The other Volkswagen commercial is the Transporter which has its origins in the VW Beetle. But the Transporter of today has improved that concept enormously, and the advent of water cooling to the still flat-four engine has made it smooth and quiet. The rear engine still means a high loading height however.*

potential is there but both the makers and buyers are treading cautiously. This particular vehicle proves that it can work, without too many of the vices normally associated with large fwd vehicles. The steering is not too heavy and the gear change and clutch action are acceptably light. Heavier than in a good Transit, say, but far in advance of some early fwd cars.

There is one other layout worth considering; that of rear-engined, rear wheel drive. The only example now remaining is the Volkswagen Transporter, which in one form or another has been going for some 30 years. The Type 2 models, to give them their generic title, were derived directly from the VW Beetle saloons. As such, their behaviour in crosswinds and on cornering left a great deal to be desired in terms of stability and safety as well as comfort. Given time, however, almost any layout can be refined to the point where it is acceptable, and the Transporter in its latest water-cooled version has achieved this to the extent that, it could be argued, had it been like that to start with, it would never have become unpopular.

The disadvantage with this particular rear wheel driven arrangement

Light Van Selection

Figure 5.8 *Owing its origins to an American design, the Dodge 50 series offers a strong truck-type chassis for top weight applications.*

is that however flat the engine can be made (the VW engine is only 22 inches high), its position means that the rear loading height is too high and makes the lifting of heavy packages awkward. In the case of the Transporter this dimension is 825mm (32.5in) which is, for example, some 325mm higher than the Renault Trafic in front wheel drive form. To be fair, the loading height of the VW is only a fraction higher than the Mercedes 407 range which is of conventional format. The VW does have one advantage which it shares with the front wheel drive vans: a low floor in the centre section. This gives the side loading aperture of the VW an excellent loading height of only 460mm (18in), which is lower than almost any other vehicle currently available. Therefore, if your operation does not need heavy loads to go in at the back, the VW could be worth considering. It is certainly an example of just how far a very basic design can be taken in terms of comfort and practicability.

The behaviour of the VW on a wet, windy motorway still cannot match the remarkable directional stability of the same maker's heavier front engined LT vans, or the front wheel driven Talbot Express, but it is still safe and responsive. The rest of the vehicle is refined, easy to drive, and quiet with this last point illustrating another advantage of rear-engined vehicles – all the noise is behind you.

When vans first stopped having bonnets and began to have forward control, many thought that it was unsafe sitting so close to the potential impact area. This is a fallacy, as many safety studies have shown. Good design can keep the driver area free of impact damage wherever the engine is located. With the VW there is nothing but sheet metal in front of the driver, but in tests at the VW track at Wolfsburg, the Transporter in a simulated collision with a car suffered minimal damage

This VW van, like the LT, now has a five speed gearbox option. The addition of an extra gear in van specifications has surprised many people. Obstinate operators look upon it as one more component to go wrong, and more than a few drivers would rather have four gears that are well matched and easy to change than have five that are not. Certainly if the location of the lever or the linkage is poor, then the addition of another gear will compound the problem, but if the design is right and the ratios are well chosen there will be less wear and tear on both driver and vehicle.

Today most journeys involve at least some motorway work, even if the operation is mainly urban. Ask anyone who has driven a one ton van at maximum gross weight on an even slightly hilly motorway whether there are enough gears. The 'slog' or 'scream' techniques can be very tiring on the nerves. At the maximum legal speed (70mph for vans up to three tons unladen weight) an engine revving at 5,000rpm in an empty steel box can be extremely noisy as well as causing excessive wear on the engine. The other side to this argument is that with gearing for relaxed motorway use becoming more common the maximum attainable speed of vans is rising. When Ford first brought out the Transit, it astonished the motoring fraternity by being indecently fast – some versions could achieve almost 80mph. Today, almost all vans can reach this speed, with some, like the latest six cylinder VW LT, being capable of 90mph.

It is not yet possible to get this level of performance from a diesel engine, which has an unjustifiable reputation for lack of acceleration in many quarters. The habit of manufacturers of installing undersized and underpowered diesel units into what were quite reasonable vans is a matter of history. The reputation persists although the current crop of engines is slowly changing the poor image of the diesel. The initial cost of the diesel engine is much higher, to the extent that it makes economic sense only if the annual mileage is very high. It can take at least 30,000, sometimes 50,000 miles before the extra cost and maintenance load is repaid with lower fuel costs, and even that is subject to the whims of governmental excise duty policy, which can change almost overnight. But the attraction of 50mpg around town, as is claimed for the latest small Ford diesel, is difficult to ignore. This figure has been quoted for the two car-derived Ford vans, the Escort and Fiesta, where the sales battle is hotting up.

Bedford has a good contender in this slot with a 1,600cc diesel in the Astra, which attains about 43mpg. These latest car type diesels follow the trend started by the VW Golf engine: light, free revving and relatively smooth, with starting difficulties now a thing of the past. They work well enough in the light van application where duty cycles are not too severe, but in certain cases of car type diesels and in the larger panel vans, noise and lack of power can still cause problems. This does not apply to the latest six cylinder diesels from Volkswagen, which are powerful, smooth and quiet in the LT model.

The LT van is also available in its heavier versions with a turbocharged diesel, and, as has happened in trucks and now cars, seems set to change the unacceptable face of the diesel engine.

Turbocharging adds power, reduces noise, improves economy, lessens smoke and improves the concept all round. It also increases cost and complication, but the effects are dramatic. If only the designers could get rid of the low speed knock and rattle of a diesel, and do away with that infamous 'wait 20 seconds' glow plug, we really might be getting somewhere. That start delay has come down, slowly, to around ten seconds in some engines, but it would be far better if it could be eliminated. Some of the latest diesel electronic injection systems suggest that this is soon going to be possible. Meanwhile, the diesel option remains roughly where it has always been – unpopular except for specific operations. However, when Bedford made the diesel premium very small – less than £400 – and put it in a good vehicle – the Astra – then even Bedford was surprised when the diesel option accounted for almost 50 per cent of total sales. The prediction was for about 80 per cent to be petrol-engined. This market seems to have now settled at about 70/30 in favour of petrol but it does show an interesting trend.

There are other power choices besides diesel and petrol although their extensive use in vans appears doubtful. Both liquefied petroleum gas and electric power have made some inroads, notably in public authority use, but, as with diesel, their long-term future is very dependent on government fiscal policy. Both are also subject to the natural and largely unjustified suspicion of potential users. The most usual charge labelled against LPG is that 'it must be dangerous carrying around all that compressed gas'. The only answer to this is the logical one that the LPG suppliers use. If you were today to try to get a patent for an engine that ran on a fuel that had a flashpoint at about room temperature and whose fumes were heavier than air so that it could drift along at ground level until it found a point of ignition, it would be interesting to see if such a design would be approved. Yet that is just how petrol behaves. The LPG container is a tested pressure vessel, not just a metal container and, as such, it is possible that it could be even safer in a crash. But, because of the current problem with poor availability of LPG, most installations are dual-fuel systems with petrol being used for starting and as a reserve. So, in the event of a crash, having a petrol fire just under a gas cylinder would be disastrous. The counter argument to this is that very few vehicle accidents involve fire of any kind.

The case for LPG on the grounds of cost is sound, provided that it remains substantially cheaper than petrol. The present difference in price is due almost entirely to the lower excise duty which it bears. However, should the price of LPG rise then this cost advantage would be quickly eroded, as the calorific (energy producing) value of LPG is lower than that of petrol, giving an increase of approximately 15 per cent in fuel consumption. Two additional advantages of LPG are that it contains no lead and that it is very difficult to steal. This second point can be an extremely significant cost factor in petrol-engined light vehicle fleets, where illegal siphoning is a common problem.

The case for electric power is much more difficult to argue. In a local delivery van application, the range and performance parameters are now just about on a par with the duty demand. Since vans are a major user of petroleum fuels – about 11 per cent of all automotive fuel consumed in the UK – the market for electric vehicles would appear to be wide open. That it is not is because of the relative expense of such vehicles and because of the low payload to gross weight ratio.

Electric vehicles are quiet (indeed so quiet as to be a safety hazard in some places), vibration-free and very easy to drive. The electric vehicle is not quite pollution free, but almost so. In theory, at least, all that is produced as pollutants will be hydrogen and oxygen, which recombine in the battery to produce water in the best system, and a certain rather small quantity of sulphurous fumes.

There are currently about 300 electric vehicles of the non-milk float type in service, with Bedford, Freight Rover and Dodge all committed to maintaining production levels in order to meet the present demand for such vehicles, which is still low. In the Far East, however, the Japanese electric vehicle industry has réally taken off and manufactures literally thousands of such vehicles, giving an indication of the real potential for this type of transport. Perhaps hybrid vehicles (part petrol, part electric) would make better economic sense for the UK market. The real reason the electric vehicle has such a limited market here is purely because of its high price. The long-term running costs might well work out as less than petrol, and the equivalent of 80mpg is very tempting. But would you pay three times the cost to obtain this figure? Unfortunately for electric vehicle manufacturers most transport operators are more concerned about paying their monthly bills than worrying about the cost savings over several years.

Four wheel drive is another specialist area that alternately achieves and then wanes in popularity. In the case of vans and pickups the answer is simple: if you need fwd all of the time, then buy a van which has this facility. If your need is sporadic, hire one instead. The extra cost and complexity of four wheel drive could cost you dear, in terms of fuel consumption and maintenance as well as in its initial cost. But if a lost delivery at the end of a muddy farm track could cost you the contract, then there is nothing like four wheel drive to get through. But remember that in many conditions a good driver in a standard two wheel drive vehicle having a relatively high ground clearance can do nearly as well in all but the worst conditions.

Back at the mundane end of the van market there is one thing more urgently needed than any new magic fuel or Quattro-like performance. It is a set of standards for van specifications. There is no earthly reason why van makers should be exempt from having to quote laden fuel consumption figures achieved in a standard test. The elements that go into or are left out of kerb weight and payload calculations should be fixed. And the figures for cubic carrying capacity which make use of such devices or tricks as shaped boxes or table-tennis balls in order to maximise this figure should be outlawed. This might go some way to making comparison charts truly comparative, so you could then judge for yourself whether X loadlength and Y cubic feet at Z pounds represent good value with a consumption of A mpg and a spares and maintenance cost of B pounds per mile. At the moment it is practically guesswork. What you *can* do now is to list all the vans available in the category you are interested in, but list them in order of deck dimension, or cubic capacity, or mpg, or whatever seems to be the most important factor in your present and future business needs. Then, delete all those that come from a manufacturer whose products have proved unsatisfactory in the past for whatever reasons. Next, cross out all the ones you just do not like, whether judged by personal experience or rumour. The chances are that there will be only one or two possible choices left.

It's easy...

To increase payload with no increase in height...
(Use a stepped body)

To obviate high stress concentration in chassis flanges...
(mount suspension direct to cross members)

To ensure utmost rigidity and stability when tipping...
(use rearmost crossmember as a combined suspension/hinge pivot, so transferring load direct to axles and suspension)

To increase driver/operator safety and convenience..
(fit heavy duty catwalk/toolbox unit)

Hoynor 40 cy. yd. Step Frame Tipping Trailer

**Available as tandem or tri-axle units.
40 cubic yard to 80 cubic yard capacity.**

when you know how!

HOYNOR TRAILERS

MANUFACTURED BY
REDMENT ENGINEERING LTD
SPRINGWOOD Industrial Estate   Rayne Road   Braintree   Essex
Tel. Braintree (0376) 21900

6. The Initial Effects of the Gross Weight Increase

Graham Montgomerie

On 1 May 1983 a new gross weight limit of 38 tonnes came into operation in the UK. This was the eventual watered-down result of the Armitage Committee's Report which was published in December 1980. Although the Report suggested a number of weight increases, including 34 tonnes on four axles and 44 tonnes on six axles, the final limit was set at 38 tonnes on five axles by means of an amendment to the existing regulations or, to give it its full title: The Motor Vehicles (Construction and Use) (Amendment) (No.7) Regulations 1982. This Amendment was laid before Parliament on 12 November 1982 before coming into operation, as already mentioned, on 1 May 1983.

As some time has now elapsed since that date, it is possible to take a look at the Act's effect on the heavy vehicle end of the transport scene to see just how operators are going about achieving the most efficient method of running at the new gross weight.

In essence, Amendment No. 7 to the Construction and Use Regulations permits a gross weight of 38 tonnes on five axles while retaining the existing limit of 32.5 tonnes (32 tons) for four axle outfits. Anything operating at above 32.5 tonnes still requires five axles whether or not it comes anywhere near the 38 tonnes maximum.

On the dimensional side, the new maximum permissible length for an articulated vehicle has now been set at 15.5m (50ft 10in) instead of the previous 15.0m (49ft 2.6in). To ensure that this extra 500mm is not merely incorporated into the platform length, the length of the load carrying space is now defined (for the first time) at 12.2m (40ft). For the purposes of this regulation, the maximum dimension excludes the thickness of the front and rear walls, headboard thickness, refrigeration unit and so on.

Under the 1968 Construction and Use Regulations the critical dimension for an articulated combination was the inner axle spacing, ie the distance from the drive axle of the tractive unit to the leading axle on the bogie. The Department of Transport argued that this rule was quite complex in practice at up to 32 tons and would become unwieldly if extended to cover the various options up to 38 tonnes. Furthermore, it was suggested that the critical spacing rule, employed to minimise the effects of heavy loads on short span bridges, was that between a tractive unit's rear axle and the *rear* axle of its semi trailer. Thus, the ruling on the axle spacing definition was changed and the term 'relevant axle spacing' was added to the new legislation covered by Amendment No. 7 to the Construction and Use Regulations which came into effect on 1 May 1983. The full dimensional requirements for relevant axle spacing are shown in the Appendix on pages 185-94.

Permitted maximum weights for three axle tractive units		
Intermediate axle weight (kg)	Outer axle spread (m)	Maximum gross vehicle weight (kg)
not exceeding 8,390	at least 3.0	20,330
not exceeding 8,640	at least 3.8	22,360
not exceeding 10,500	at least 4.0	22,500
not exceeding 9,150	at least 4.3	24,390
not exceeding 10,500	at least 4.9	24,390

For a tri-axle trailer, the minimum distance between adjacent axles required to achieve the full 38 tonnes gross weight is set at 1.35m (4ft 5in) giving a bogie spread of 2.7m (8ft 10in). This allows a maximum individual axle weight of 7.5 tonnes and thus a maximum bogie load of 22.5 tonnes. The regulations also make provision for a sliding scale of axle weights and spacings down to 6.0 tonnes at an individual spacing of 0.7m (2ft 3.6in).

As mentioned earlier, for a vehicle of 38 tonnes gross weight a minimum of five axles is required with no legal ruling as to where this 'extra' axle should be placed. The immediate reaction from most operators was that a tri-axle trailer was a logical point at which to start and this caused a boom in the trailer market, both for new products and for conversions to existing trailers.

Converting existing vehicles is the cheapest way – in the short term – to achieve the required minimum number of axles to permit the hauling of 38 tonnes, as few operators can afford to buy new equipment immediately. It also allows an operator a period of breathing space in which to judge whether the 3 + 2 or 2 + 3 configuration is right for his particular operation.

The quickest method of conversion is to add an extra axle to the trailer, using tyres which match the original specification. The majority of tandems in current use are fitted with 11 × 22.5 tyres and adding an extra axle to match the existing twin wheel specification costs approximately £2,400. This is very much a 'ball park' figure as it depends not

The Initial Effects of the Gross Weight Increase

Figure 6.1 *The maximum length of the trailer has been defined for the first time. It has been set at 12.2m (40ft) which excludes the thickness of the front and rear walls.*

only upon the conversion company but also on the operator providing his own extra axle.

In most cases, the existing suspension has to be relocated. It is possible to reach a satisfactory load distribution merely by adding the extra axle in front of the existing bogie but such cases are rare. Moving the complete suspension to achieve the right distribution can cost up to £300 extra.

One of the attractions of this type of conversion is that those operators having surplus trailers, as a result perhaps of the economic recession, can convert three tandems into two tri-axles by simply splitting an existing bogie and adding an extra axle to the other trailers. Admittedly, this leaves the operator with a spare frame but it is still an option worth considering. The main thing to remember about trailer conversion costs is that the quoted price will vary considerably depending upon how much work is required and how many new components are needed.

A more permanent and long-term method is to replace the existing twin-wheeled tandem bogie with three axles using single wheels and 15 × 22.5 tyres. This obviously involves the total replacement of the

Figure 6.2 *To achieve the maximum bogie load of 22.5tonnes, the minimum distance between adjacent axles is 1.35m (4ft 5in).*

existing equipment and so the cost is higher, with £3,600 being a typical figure if the axles, wheels and tyres are included. The stability of the big, single tyred outfit is better as it is a case of going for the widest track possible. A typical twin wheel arrangement has a 965mm (38in) frame centre compared with the 1,143mm (45in) of a new build tri-axle on single wheels. Changing from a twin to a single wheel specification also brings a weight saving advantage. Compared with twin wheels, the big single saves about 400kg (8cwt). A further option for trailers is to have a self-tracking back axle, but this can usually only be added to the narrow type of frame as the 45in frame does not permit enough steering movement.

An important factor to be considered when adding an extra axle to the trailer is that of braking efficiency. The change in axle weight from the 10 tons per axle of the tandem to the 7.5 tonnes of the tri-axle means that the braking efficiency has to be reassessed. If the existing braking equipment was left untouched then the 7.5 tonne axles would be vastly over-braked especially when unladen. Thus, in almost every case the trailer will end up with six new brake chambers.

A reputable trailer conversion specialist must make sure that a trailer conversion meets the new legal requirements on axle spacing and minimum overall length and this can cause problems with trailers shorter than 9.7m (32ft). The slope of the trailer needs to be reassessed for

Leaders in specialised transport technology.

Rugged, reliable Scania trucks, immaculate Saab turbo cars, the advanced Saab-Fairchild 340 airliner. They are all products of the international Saab-Scania organisation, and represent just a few examples of how Saab-Scania aims for excellence through concentration and specialisation. For further information, contact: Scania (Great Britain) Limited, Tongwell, Milton Keynes MK15 8HB, Buckinghamshire. Tel: 0908 614040. Telex: 825376.

Figure 6.3 *Trailers shorter than 9.7m (32ft) can often be difficult to convert to a tri-axle configuration as the slope of the trailer needs to be reassessed to permit proper load equalisation.*

proper load equalisation as the equaliser must be free to operate correctly. The situation can arise where the rocker can get bound to the underside of the trailer so axle heights relative to the chassis are critical.

In many cases, the structure of the trailer needs attention to enable the final tri-axle to be used at 38 tonnes gcw. With the original tandem specification this implies strengthening the front end. This means moving the kingpin rearwards and reinforcing the neck to enable it to accept an imposed load of some 13 tonnes. In turn, if the kingpin needs to be moved then the loading legs need to be moved as well. The frame itself might need to be reinforced to cope with the extra scrub loading generated by the tyres on a tri-axle bogie. According to Crane Fruehauf the side load generated in a tight corner on a dry road surface equates to the static axle load.

Although these conversion comments have applied to flat platform trailers it is possible to convert frameless vans and tippers, but in these cases additional constraints apply. With the former a subframe extension is required, while for refrigerated transport the weight of the fridge unit itself must be taken into account before deciding on the new position for the bogie.

With tippers a lot of frame reinforcing is usually required, as is stiffening of the beams. Converting a tandem tipper to a tri-axle is straightforward enough, but it is difficult to convert an existing tandem to run at 38 tonnes with a three axle tractive unit. The centre of gravity must be moved forwards, which necessitates moving the body and ram forward

Figure 6.4 *Having the third trailer axle suspended by air is a popular option whether or not a lifting facility is incorporated. This is the York Airpoise conversion to existing steel-sprung tandem trailers.*

Made for low-cost distribution.

Now, a choice of three demountable systems for low-cost distribution with a full range of bodywork options.

The Alcan Bonallack Loadspeeder range has been extended to give you a choice of 3 demountable systems; hydraulic, mechanical or the new Hydrail® system – all in the full range of vehicle sizes, with increased payloads.

For all the details of demountable systems and expert in-house body-building for them, contact Colin Sykes:

Alcan Transport Products Ltd.
Alverthorpe Wakefield Yorks WS2 0AQ Tel: (0924) 371141.

as well, as otherwise a bending movement would result between the ram and the fifth wheel.

Having the third trailer axle suspended by air is becoming a very popular option, although the legality of this approach was in doubt until very recently because of the requirements for load compensation. Regulation 11 of the 1978 Construction and Use Regulations states that:

> Every motor vehicle or trailer with more than four wheels and every trailer having more than two wheels being part of an articulated vehicle shall be provided with such compensating arrangement as will ensure that all the wheels will remain in contact with the road surface and under the most adverse conditions will not be subject to abnormal variations of load.

The problem arose with the definition of 'such compensating arrangement'. The official view of the Department of Transport is that there is no reason why a 'mixed' suspension such as air and metal leaf should not be capable of satisfying the requirements of Regulation 11. In principle, the DoT suggests that this is possible if the load imposed on the bogie is shared in a pre-set ratio between the wheels of that bogie, and that this load sharing should be proportionally applied between the laden and the unladen condition.

The Granning system is typical of an 'add-on' air suspension and

Rubber and Air Suspension Systems for Rigid Vehicles and Trailer Running Gears

Singles   10-13 tonnes
Tandems   18-40 tonnes
Triaxles and Third Axles to suit new legislation
- Long life, low maintenance
- Premium quality, bespoke engineering
- 24 month, 250,000km warranty
- High articulation, fail safe design
- Low deck height ability
- 'Specials' customer engineering capacity

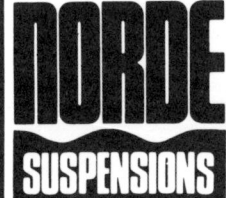

NORDE Suspensions Ltd.

Sywell Airport, Northampton NN6 0BU England
Telephone: Northampton (0604) 493161
Telex: 311400

---

costs around £2,700, excluding the wheels and tyres and excluding any repositioning of the existing bogie. The weight penalty is in the region of 900kg (18cwt).

From the point of view of cost, it might appear that the obvious way to run at 38 tonnes would be with an existing 4 × 2 tractive unit and a tri-axle trailer. But it is not quite as simple as that.

At 32 tons gcw the drive axle load limit was 10 Imperial tons or 10,170kg. When the gross weight limit was raised to 38 tonnes, however, this axle limit was increased to only 10,500kg and this has brought the risk of overloading the drive axle even with a purpose-built tri-axle trailer.

As part of the overall environmental 'package', wrapped up with the original Armitage proposals, and the resultant legislative changes, the penalties for overloading were increased considerably. Previously the fine was £400, but this figure is now being set at £1,000, and it must be stressed that this fine is for *each infringement*. In other words, if the outfit is found to be exceeding not only the gross weight limit but also the individual axle weight on, say, three axles, then the fine is £4,000 in total. Thus, the incentive to stay within the loading requirements is a very real one, but again this is more difficult than it sounds to put into practice.

One popular misconception is that all trailers have a uniformly dis-

Figure 6.5 *For operators with a potential drive axle overload problem, the three axle tractive unit offers more flexibility.*

tributed load, which might be feasible for an own account operator, but it is unlikely to be achieved in general haulage. Even in the case of the former, a diminishing load type of operation renders such a weight distribution impossible. For operators engaged in hauling containers or loose trailers this problem is magnified as they have little or no control over the way the trailer is loaded. For those operators the danger of overloading is such that most are preferring the other 38 tonne option, the three axle tractive unit hauling a tandem trailer, as giving them greater flexibility of operation without the inherent dangers of potential heavy fines.

Because of their use of higher gross weight limits for many years, continental truck manufacturers have had a lot of experience in building 6 × 2 and 6 × 4 tractive units, although the latter have been intended mainly for heavy haulage. Thus, Mercedes-Benz, DAF and Iveco were very quick to offer 6 × 2 tractive units for the UK market, to be followed within a few months by Scania. Volvo was rather a different case as this particular 6 × 2 was a conversion carried out by the Scottish subsidiary at Irvine, rather than a definite Gothenburg model.

Representing the UK manufacturers, ERF was way ahead of the others in having a purpose built 6 × 2 tractive unit, actually exhibiting the first example at the Birmingham motor show in October 1982. Since then, Seddon Atkinson and Leyland have released their equivalent models while Foden Trucks has preferred to follow the double-drive 6 × 4 route. In most – but not all – cases the manufacturers have gone for a 6 × 2 tractive unit with the second axle having a steering capability. This has lead to some confusion over definition as the terms

The Initial Effects of the Gross Weight Increase

Figure 6.6 *DAF is one of the continental manufacturers to offer a rear steer tractive unit for use at 38 tonnes gcw.*

'twin steer' and 'rear steer' have both been applied. In all cases, however, the position of the second axle is immediately in front of the drive axle rather than immediately to the rear of the front steering axle.

Of the 6 × 2 tractive units currently available, the Swedish models vary from what can be considered the norm, in that the Scania has a trailing, non-steered third axle and the Volvo extra axle is self-tracking rather than part of the connected steering linkage.

Volvo was quick to produce a 6 × 2 for the UK market but argued that a positively steered second axle was unnecessary, opting instead for a self-steering axle. This avoids the packaging problems of the extra steering linkage and saves an estimated 150kg (300cwt) in weight. As mentioned earlier, the 6 × 2 Volvo is a conversion carried out by the modified vehicle assembly (MVA) department at the company's Scottish factory at Irvine. The longest wheelbase in the Volvo 4 × 2 tractive unit range is 3.8m (12ft 5.6in), which is the minimum required for a plated weight of 22.36 tonnes, so virtually no chassis modifications are required other than the addition of the second axle. Air suspension is used for the second axle while retaining parabolic leaf springs for the front and drive axles. The self-steering axle is a single tyred standard Volvo front axle with increased caster to allow it to

Figure 6.7 *The Leyland 20.32 is unusual in that it does not use air suspension for the second steering axle.*

track. For reversing, the axle is locked in the straight-ahead position using air pressure which is activated as soon as the driver selects reverse gear. The same principle of operation applies when the forward speed exceeds 60km/h (38mph).

Leyland is one manufacturer which has decided that, in practice, no six wheeler operating in the UK is likely to exceed 20 tonnes gvw and accordingly has gone for a 3.45m (11ft 4in) oas limiting the new Roadtrain 20.32 to 20.33 tonnes gvw. This specification was chosen to allow maximum trailer interchangeability, to the extent that Leyland is claiming that 12.2m trailers with kingpin positions between 760 and 1,120mm (30 and 44in) will couple at full imposed load and within the legal overall length limit when the fifth wheel is positioned 700mm (27.6in) ahead of the drive axle. Alternatively, trailers with kingpin positions between 865 and 1,220mm (34 and 48in) will couple with the fifth wheel at 600mm (23.6in) with no loss in imposed load capacity. This specification allows Leyland to claim that the Roadtrain 20.32 is interchangeable with most of the 4 × 2 tractive units in existing fleets.

Whereas most of the opposition have opted for air suspension for the second steering axle, Leyland uses a single leaf spring which is slipper ended at the rear and shackled at the front. Unladen, the trailing end of

THERE'S ONE VAN YOU CAN RELY ON FOR TWENTY YEARS.

In 1963 Frank Willis and Sons (Carriers) Limited, one of the founder members of United Parcels, bought their first York Freightmaster. 20 years on, it's still operating.

As you would expect, the 1983 Freightmaster has many new features designed to improve the van's reliability, like corrosion-resistant cadmium-plated base panel fixings, and the latest impact-resistant rear end. Features that operators value in the battle to minimise operating costs.

The laminated hardwood floor over 'I' beams at 12" centres takes a fork lift truck the full length of the trailer.

And an all-steel welded base frame means that there are no fixings to work loose during its years of service.

York have set the standard by which good van design is measured.

If, like United Parcels, your business depends on offering a first-class and totally reliable service to your customers, then take a look at Freightmaster.

There is no van with the same reputation for reliability and durability.

And without such a past, who else can guarantee such a future?

York Trailer Company Limited, St. Marks Road, Corby, Northants. Tel: 05363 3561. Telex: 34516.

DIAL 100 AND ASK FOR FREEFONE YORK.

TALK TO YORK. UNITED PARCELS DID.

York Freightmaster. You can't cut costs by cutting corners.

103

the spring can float vertically through some four inches, which means that the second axle suspension carries only the weight of the axle. Once a force of some 4,000kg has been applied to the fifth wheel, the conventional springs on the first and third axles deflect, thus lowering the chassis and bringing the second axle suspension into action. This second axle carries 4,730kg while the first and third axles can take 6,500 and 10,500kg respectively. The kerb weight of the sleeper cab Roadtrain 20.32 is 7,630kg (7tons 10cwt) which includes fifth wheel, ramps and a full (377 litres or 84gal) tank of fuel. The power unit is the turbocharged and charge-cooled NTE 320 Cummins which produces 227kW (304bhp) and 1,430Nm (1,056lb/ft) torque.

ERF offers two three-axle tractive units, the twin steer already mentioned and a 6 × 4 option. The twin steer is available with the usual ERF mix of Cummins, Gardner and Rolls-Royce engine options, which obviously results in a variation in kerb weights as does the sleeper cab option but, as a guide, a Rolls-Royce 290L powered ERF in day cab form weighs 7,015kg (6tons 18cwt) when fully fuelled.

This particular ERF utilises a novel bogie arrangement with rubber suspension units which are installed lengthways. Lateral location is provided by an A-bracket attached to the drive axle. A rocking beam is pivoted to give the correct ratio of 10:6 for the axle loadings which, on the ERF, are set at 6,100kg for the steered axle and 10,170kg for the drive axle.

The 6 × 4 ERF uses a lightweight, leaf-sprung version of the Hendrickson bogie which incorporates a balance beam made of aluminium. Longitudinal location is taken care of by radius rods. The length of the eight cylinder 8LXCT engine makes it impossible to install in the 6 × 4 ERF but the other engine options stay the same. Using the same 290L engined version as an example, the 6 × 4 ERF has a kerb weight of 7,165kg (7tons 1 cwt).

Foden Trucks has opted for a 6 × 4 configuration for its Foden six-wheeler and is unusual amongst its rivals as this is instead of a 6 × 2, not as well as. As with its rival ERF, rubber is used for the bogie suspension, in this case the Foden FF bogie which the company has offered on its multi-wheeled rigids for some time now. Although the rubber suspension is as near a 'standard' component as it is possible to get within the Paccar concept of vehicle building, Foden does offer an alternative suspension in the form of the Kenworth torsion bar system.

The main advantage claimed for this system over a leaf spring suspension is that it spreads the load over more of the chassis. The torsion bars are nearly 2.5m long with one bar at each side of the bogie. The Foden uses a Rockwell bogie with the individual type depending upon operator requirements. Extensive use of aluminium keeps the kerb

The Initial Effects of the Gross Weight Increase

Figure 6.8 *Whereas most manufacturers have opted for a 6 × 2 configuration, Foden Trucks prefer a double-drive bogie. Extensive use of aluminium helps to keep the weight down.*

weight down to the extent that a Cummins L10 engined version with torsion bar suspension and with front wheels, air and fuel tanks in aluminium weighs 5,925kg (5 tons 19cwt).

Foden argues that traction can be a problem at 38 tonnes with a maximum of only 10.5 tonnes on the drive axle on a dry road. If the road is wet and the vehicle unladen then traction problems are likely on any gradient steeper than, say, 12 per cent or about 1 in 8. Although a double-drive helps with traction it can cause handling difficulties because of the tendency of the vehicle to go straight on at corners. Because of this handling characteristic – and it is a characteristic of double-drive rather than a serious problem – front tyre wear can be high.

Mercedes-Benz has two rear-steer tractive units available: the 2025S and the 2028S. Apart from a difference in power output, the major difference is in the suspension department. The 2025S uses air suspension for the second steering axle only, and four-leaf parabolic springs for the drive axle, whereas the 2028S uses air suspension for both. Mechanically the 2025S follows closely the specification of the 4 × 2 1625S which includes a turbocharged OM422 Mercedes V8 engine which produces 185kW (247bhp) at 2,300rpm with a maximum torque of 932Nm (687lbf) at 1,200rpm. A 16 speed Ecosplit gearbox from ZF is standard.

Figure 6.9 *The 2025S from Mercedes-Benz has air suspension on the second steering axle only in contrast to the higher powered 2028S which also uses air for the drive axle.*

Apart from having the choice of a number of brand new three axle tractive units announced since the raising of the weight limit, the operator can also follow the conversion route with his existing machines. York and Primrose, for example, have been adding a third axle for many years to convert a 4 × 2 into a 6 × 2 and this business has received an added impetus because of Amendment 7.

Although several vehicle manufacturers and conversion specialists have claimed that their three axle vehicles will allow coupling up to any of the existing trailers in a fleet, this sort of statement needs careful analysis, as in most cases it is totally misleading. The problem with picking up existing tandem trailers is that the neck profiles and landing legs can foul – particularly on conversions with a third axle added behind the drive axle – so each individual case needs looking at very carefully.

The biggest challenge to the 3 + 2 combination is to be able to couple up within the new 15.5m overall length limit. This is much harder with a six-wheeler than with a 4 × 2 tractive unit and, with the trailing axle vehicle especially, the fifth wheel needs to be biased forwards, which makes the neck clearance problem worse. These clearance comments

The Initial Effects of the Gross Weight Increase

Figure 6.10 *As well as having a number of new 6 × 2 tractive units to choose from, the operator can have his existing 4 × 2 converted as in the case of this Scania 112 which had a second steering axle added by York.*

apply where a three axle tractive unit is being used with an existing tandem trailer. A tandem designed primarily for 38 tonnes gcw should have no difficulties in this respect.

The extra axle can also go in front of the drive axle but in this case it is a steered (or at least self-tracking) axle. With the second steered axle close to the drive axle the fifth wheel is biased towards the rearmost axle, so immediately there is a better chance of coupling up to existing trailers without fouling the landing legs.

There are several conversion specialists in the tractive unit field although some concentrate on specific makes only. ERF, not surprisingly, will only convert ERF chassis while Lyka has received official approval from Seddon Atkinson for its conversion to the 400 and 401 tractive units.

One point to watch with a tractive unit conversion is that there are arguments both for and against opting for the maximum outer axle spread. In Amendment 7 to the Construction and Use Regulations, a 3.8m (12ft 5.6in) oas gives what is considered the maximum practicable gross weight of 22.36 tonnes, but in many cases this gives rise to serious coupling problems. With conversions it is often a more attractive proposition to go for a shorter oas – providing this exceeds 3.0m (9ft 10in) – and accepting a gross weight limitation of 20.33 tonnes. The same argument can of course be applied to new vehicles and MAN for one has opted for a 3.6m (11ft 10in) outer axle spread.

It is almost impossible to quote a typical conversion cost for turning a

4 × 2 into a 6 × 2 as so much depends on the make of the vehicle. Often a chassis extension is required and this is accompanied by relocation of the pipework and the fuel and air tanks. To be considered very much as an approximate guide, York quotes around £5,500 for adding a second steering axle, which adds about 1,100kg (22cwt) to the kerb weight.

It is also impossible to make a definite recommendation as to which is the best route to follow for a 38 tonne combination as there are so many different variables to consider from operation to operation. The cheapest, as far as initial cost is concerned, is a 4 × 2 tractive unit plus tri-axle trailer, but this combination is taxed at a higher rate and has little or no flexibility on axle loading. The three axle tractive unit is more expensive to buy but cheaper to tax. It gives far more loading tolerance but can be difficult to couple with existing trailers within 15.5m overall.

In conclusion, an operator should be very certain of his operating requirements before he is persuaded to follow any particular path to 38 tonnes.

7. The World Truck Concept

Graham Montgomerie

Truck building is a very expensive business. In 1983 it was estimated that to take a new truck range from the drawing board to the market place would cost in the region of £200 million and to build the factory for the assembly of that range would add a further £150 million. These are not guesswork figures but the results of a carefully costed analysis carried out by one of Europe's largest commercial vehicle manufacturers.

This costing becomes potentially much higher when one considers that a truck range designed for Europe is not necessarily ideally suited for markets in Australia, Africa or the USA. This is a conclusion which the car industry arrived at a long time ago, with the result that the major manufacturers have tended to produce what could be described as a compromise design which, with minor adaptions to suit local requirements, could be sold in any market in the world. A 'World Car' in fact. Some of the Japanese models are typical, with the major components being identical in specification wherever the car is being sold.

The advantages of the concept are obvious. First, having one car for the world market is far cheaper than needing five or six models all aimed at different markets and, secondly, economies of scale increase accordingly. The manufacturer can thus concentrate on one bodyshell for example, rather than tooling up for five or six different models. The idea is working reasonably well in the car sector with possibly the most famous example being the General Motors J-car, sold in the UK as the Vauxhall Cavalier. Thus the World Car might be close to reality, but could the concept be extended to produce a 'World Truck'?

Whereas great economies of scale are possible at car factories, it should not be forgotten that what is considered 'high volume' for the truck industry is about two days' output by Detroit standards. It appears to many people in the commercial vehicle industry that the only way for many companies to survive, let alone build a World Truck, is by collaboration on joint projects, or even merger. Several vehicles and component manufacturers have already become linked in some form or another to share production costs, and so the World Truck idea comes closer if only for reasons of mutual survival.

At this point it is worth looking at the structure of some of the major

Figure 7.1 *Whatever the individual's opinion of the World Truck as a concept it is difficult to visualise designs such as this GMC Brigadier being successful in Europe for example.*

automotive groups in the commercial vehicle world to see just how far the progress of merger and collaboration has already gone.

One term which crops up again and again when the World Truck concept is discussed is that of the 'Club of Four'. This was a consortium formed in the early 1970s between DAF, Volvo, Saviem and Magirus Deutz to develop a joint truck range in the six to 13 tonne gvw category. The four member companies pooled their technical resources to design the basic vehicle and used their increased joint buying power to obtain proprietary components at reduced cost.

For example, this weight category was unknown territory for Volvo, a company famous for its trucks aimed at the heavy end of the market. To design and develop a new chassis for the middle sector would have been prohibitively expensive, but if such a cost were to be divided by four...?

The World Truck Concept

Figure 7.2 *DAF and Magirus Deutz were two of the original members of the Club of Four as can be seen in the similarity between the two cabs* (see Figure 7.3).

Figure 7.3 *The Magirus Deutz cab.*

The membership of the consortium came about partly as a result of existing cooperation on the passenger car manufacturing side. Saviem represented the commercial vehicle division of Renault, and Renault was supplying the petrol engines for the DAF Variomatic (since sold to Volvo) passenger car. Renault and Volvo were also involved in joint discussions over car interests, so from there to commercial vehicle cooperation was but a short step. Although popularly known as the Club of Four, the company was officially known as the Société Européène de Travaux et de Developpement (ETD) and was under the chairmanship of Maurice Bosquet, the deputy director general of Saviem.

Following the signing of the cooperation agreement, a group design office was set up near Paris in 1972 with a staff of about 40, the great majority of whom were 'donated' by the member companies. Once the design expertise had been pooled and the basic parameters drawn up, then the second function of the ETD came into operation. This was the policy of bulk buying of the most suitable component at an advantageous price, due to the combined buying 'muscle' of four companies instead of one. A case in point was the gearbox, which was a five-speed all synchromesh unit from ZF, designed specifically for universal use within the Club range.

Engines were excluded from the joint agreement with each manufacturer supplying its own, whether home built as in the case of Magirus, or proprietary units (Perkins) for some of the early DAF and Volvo versions. Although the research and development, buying and production aspects were carried out as a joint venture, this cooperation did not extend to the marketing side where each member was free to sell its trucks in whatever country it chose. There was no 'carving up' of the European market with, for example, Germany being the sole preserve of Magirus Deutz.

The Club of Four still exists but it is only in the form of a buying department for component supply to the original member companies. The consortium was conceived as a temporary arrangement to provide each member with an extension of its range and that is just what it did. What complicates the Club of Four story is that two of the original four became involved with other manufacturers but in these cases it involved more than mere technical collaboration.

In January 1975 the Industrial Vehicles Corporation (IVECO) was formed, which brought together the commercial vehicle division of Fiat (which in itself included Fiat, OM and Lancia Veicoli Speciali in Italy and Unic in France) and the Magirus Deutz vehicle building section of the Klöckner-Humboldt-Deutz empire. Thus Iveco immediately had access to the Club components although, with the exception of some

The World Truck Concept

Figure 7.4 *The Club of Four cab is now also used in modified form on the Mack Midliner. Renault Vehicules Industriels acquired control of the American company in 1983.*

light to medium weight Magirus models, this has never been acted upon. (The Club cabs, however, are still fabricated at the Ulm factory of Magirus on a straightforward customer/supplier basis.

The other Club member to become involved in a merger was Saviem who joined forces with Berliet in 1975 to form Renault Vehicules Industriels. Saviem, like Volvo, had taken the basic Club cab and gone a stage further by increasing its overall width so that it could be used for heavier vehicles than originally intended. It is this cab which is used on the Renault G260 which won the 'Truck of the Year' award for 1983. The Volvo F7 cab, while it looks at first glance to be the same as that on the Renault – with minor cosmetic differences – is not in fact the identical structure. Both Saviem and Volvo took the Club cab as a basis and enlarged it separately; there was no collaboration in producing the wide version.

A further extension of the Club cab usage, and a possible further pointer to the success or failure of the World Truck concept came when RVI took effective control of Mack Trucks in the USA. In 1979 Renault took a 10 per cent stake in Mack which was increased to 20 per cent in July 1982 and rose still further to 45 per cent in June 1983.

The Mack MS 250 from the Midliner range is just one result of the

RVI connection and is available either as a rigid or tractive unit in the Class 7 weight sector. It shares the same basic mechanical specification as the more powerful MS 300 with the exception of the engine. The MS 300 is fitted with a 157kW (210bhp) 8.82 litre turbocharged engine whereas the lighter MS 250 uses the 5.5 litre turbocharged and charge-cooled power unit from the Class 6 MS 200 model. Known as the MIDR. 06.02.12, this is the same engine which is fitted to the Renault G170 16-tonner for the UK market. Thus, not only does the Club cab appear on both the Midliner range and some of the Renault models for Europe, some engines are the same as well. By late 1983 the original Club cab was used in one form or another by DAF, Volvo, Iveco, Renault and Mack.

International Harvester of the USA has made two attempts to establish a serious foothold in Europe although, it must be mentioned, with limited success. In the early 1970s IH bought a one third share in DAF Trucks in Holland who were in financial problems at that time. With the benefit of hindsight it is difficult to see just how IH expected to increase its shareholding as the remaining shares were held by the van Doorne family and by the Dutch state mining corporation. Effectively, this came to nothing and IH sold its stake in DAF in 1983. (One of the few physical, rather than financial, results of the IH/DAF connection was the marketing of a Cummins-powered DAF 2800 in South Africa under the International brand name.)

The other major IH involvement got considerably further when in 1974 the American company acquired control of Seddon Atkinson. This resulted in considerable rationalisation of components including the use of the German built International D-358 engine in the original 200 16 tonner, launched in 1975 and followed by the International RA-57 drive axle for the 401 heavy tractive units.

IH also tried to join forces with the Spanish company Empresa Nacional de Autocamiones (ENASA) with a view to the International 466 engine being built in Spain at a new factory which would be 65 per cent owned by IH. It was also suggested that the financial investment by IH in ENASA would enable the latter to make its products more competitive with the rest of Europe when Spain joins the EEC and tariff barriers are dropped. However, the financial problems of IH in the USA have meant that the ENASA venture is shelved and Seddon Atkinson is up for sale.

Not all joint European/American ventures have been unsuccessful, however, although it must be admitted that none have been without their problems. Other European incursions into the American market have included Volvo gaining control of the White Motor Corporation while Daimler-Benz took over Freightliner.

Can you account for the cost of every transport mile?

We can.

Mercedes unique transport consultancy offers free advice on every aspect of your individual transport needs.

From years of experience we can provide up-to-date information, statistics and case histories to help you run a fleet more effectively and therefore more profitably. We'll even work out the exact specification of truck to suit the job. When to buy, what to buy, and how to get the best from it.

And because we're in the business of transport and not just trucks we can actually tell you how much your truck is costing you every single mile of its working life. So when you invest in a Mercedes you do so in the knowledge that there isn't a better truck for the job.

Mercedes trucks are famous for low operating costs, low fuel consumption and less down time and they're backed by the most complete service systems a truck manufacturer could offer.

Can you afford to pay the price of any other truck?

For more information, phone Mercedes-Benz (United Kingdom) Limited, London 01-561 5252 or Wakefield (0924) 254111.

Mercedes-Benz

Meticulous engineering doesn't cost you. It pays you.

Figure 7.5 *Another European manufacturer to have an involvement in the USA is Volvo who took control of White in 1981. The F7 (again with a variant of the Club of Four cab) is seen alongside its more typically American partners.*

It seems likely that there will be a further concentration of the truck industry into a smaller number of massive companies with the obvious examples being Daimler-Benz, Iveco, Renault and Ford. The smaller companies are likely to survive only by concentrating on specific market niches, taking advantage of the lack of flexibility of the larger conglomerates. As a yardstick, it is normal to describe as 'small' a truck manufacturer with a yearly output of below 20,000 units.

Even the probable long-term survivors have been considering the possibility of joint ventures to spread the cost. Along with the Club of Four already mentioned, there has been the MAN and Volkswagen agreement to produce a range of vehicles which dovetailed between the VW light vans at one end and the heavy MAN trucks at the other, with the resultant models being sold under the marque name of MAN-VW.

Most of the joint ventures however have been in the field of components rather than of complete vehicles. Recent examples of the trend have been numerous and include Leyland and Cummins on engines, Iveco and Rockwell on drive axles, Iveco and Eaton on gearboxes and ENASA and ZF, also on gearboxes. The two joint exercises involving Iveco have been to design and develop totally new ranges of gearboxes and axles at effectively half the cost to each. The ENASA agreement is different in that the Spanish company builds components

The World Truck Concept

Figure 7.6 *The American trends in vehicle design are typified by this White Road Commander 2 chassis. Extensive use of aluminium helps to keep the kerb weight down to an ultra-competitive level.*

for ZF and assembles ZF gearboxes for use in the Spanish market.

One company which has fought shy of any joint venture is Scania. In spite of the company's other factories (Brazil for example), Scania has never been considered as producing vehicles in volumes large enough to justify staying with a vertically integrated structure. It should be remembered that Scania, like Swedish rivals Volvo, produces its own cabs, its own engines, its own frames, its own gearboxes and its own axles. Not even Daimler-Benz can claim to do this.

Although it has often been suggested that the two Swedish manufacturers would be a prime target for the proprietary component manufacturers, both have strenuously denied any such moves. The Scania argument is that the company should exercise full control over the design, development and production of all the major components just mentioned and thus package the final vehicle as being best suited to the customer's specific requirements. To this end, Scania has recently been quoted as saying that the company will never be a major customer for cabs or drive-line components. This does not mean, however, that the Sodertalje intends to shoulder the whole burden of future Scania models.

Scania has also been quoted on company policy that it is prepared to

Figure 7.7 *A further example of inter-marque collaboration was when MAN and Volkswagen joined forces to produce the MAN-VW range.*

play an active part in sharing the cost of joint development work and to bear the costs of tooling used by suppliers to produce components for Scania. The last part is significant; Scania does not intend to share the tooling for components which could be used by other manufacturers.

One significant trend towards the World Truck is the manufacture of a range of vehicles in different countries using many common components. The drive lines could be identical, and it is in the areas of suspension, frame and cab that any differences occur. Take the so-called 'Third World' countries for example. Few of these markets demand a cab with the sophistication of the modern European designs, but a workable compromise would be to have a universal cab structure which could be adapted easily to the requirements of the local conditions by altering the cab furniture and the level of the trim. Leyland, for example, already does something along these lines with the Landtrain, which uses the basic G cab structure from the Bathgate range of Terrier, Mastiff and so on.

The concept of the World Truck is unlikely to reach the production stage as there are too many variables, such as local domestic legislation, to be considered. The future structure of the commercial vehicle industry is likely to be based on the results of the joint ventures which will serve to blur the dividing lines between many companies. The nearest the industry will get to the World Truck is a common selection of major components which can then be adapted to local conditions. The real stumbling block will be the cab, as already mentioned, which will need to be produced in forward and normal control versions even before any day and sleeper cab variants are considered. For the mechanical components, however, such a commonisation is practical even if the resultant vehicles do not resemble one another from the outside.

8. Research and Development Facilities

Graham Montgomerie

Introduction

To survive in the 1980s the commercial vehicle manufacturer must have an effective product, and one which is tailored to the needs of the market place. The smaller manufacturers (whatever their country of origin) tend to concentrate on particular niches while the likes of Daimler-Benz, Iveco, BL and Renault are forced to spend massive amounts of money on research and development to come up with a product which is not only competitive but also meets any known, or anticipated, legislation in the country where the vehicles are to be marketed.

Furthermore, the larger manufacturers tend to be more vertically integrated and thus need to provide development facilities for chassis, engines, gearboxes, cabs and so on – again to satisfy the needs of different markets around the world.

The Leyland Technical Centre

One of the most advanced research facilities in the automotive industry is Leyland's technical centre which cost £22million to build and was opened in 1980. It employs 200 people and consists of three sections: a laboratory/rig test building; an extensive combination of test tracks; and a vehicle workshop which includes areas for component stripping and inspection.

The laboratory building houses a number of separate testing functions, such as the environmental chamber which enables components to be tested under extreme conditions. The temperature of the chamber, for example, can be varied between -40° to +120°C (-40° to +248°F) while the humidity can be controlled between 20 and 95 per cent. The cooling systems chamber, as its name suggests, is used for testing the engine and cooling system under actual running conditions by means of an electric dynamometer.

One of the major legislative influences on vehicle design is that of noise limitation and so it is not surprising that the Leyland technical centre includes a semi-anechoic chamber. This is an echo-free room in

Figure 8.1 *Prototype vehicles are tested for many miles before being allowed out on the roads. This Leyland Roadtrain is being tested on a Schenk road simulator.*

which analytical noise measurements can be made on complete vehicles, systems or components. The chamber incorporates a chassis dynamometer so that the noise measurements can be taken under representative running conditions. The 'chamber' is made from reinforced concrete in boxed sections and weighs no less than 1,400 tonnes. The chamber itself is 'isolated' on steel springs to reduce background noise. The walls and ceiling are covered with an absorbent lining of 5,000 glass fibre wedges, each one a metre in length. The chamber measures 19 × 9 × 5m high (62ft 4in × 29ft 6in × 16ft 5in) and can accommodate the largest size of commercial vehicle currently produced. A computer data logger is housed in a separate control room along with the noise analysis equipment.

The Leyland technical centre is also equipped with an electro-hydraulic ride simulator. This can be used to investigate ride characteristics, or for endurance testing by enabling the effect of road conditions to be accurately reproduced on a complete vehicle. Six electric pumping

Figure 8.2 *Larger manufacturers need to provide development facilities to test a large number of in-house components, as in this Renault power train test cell.*

units feed two ring-main systems, each ring-main feeding three actuators on which the vehicle is located. The equipment is activated through a computer-controlled, random signal generator, which enables a wide variety of road surfaces and vehicle speeds to be reproduced. The entire rig is isolated from the main building structure by being mounted on a 1,000 tonne concrete block mounted on air springs.

The electro-optical laboratory is a general laboratory with black-out facilities for assessing vehicle lighting, while the brake rig enables the performance of braking systems and materials to be monitored under dynamic conditions. It can be used for determining friction material performance and wear rates as well as for investigations into brake design and cooling.

The test track, part of the latest Leyland investment, covers a total of 150 acres and comprises a number of features. The outer test track is 1.9km (1.1 miles) long and is a high-speed banked circuit with a 'hands off' speed of 120km/h (68mph). It has three separate speeds at the same time. The inner 'number two' circuit is designed for low speed running

Figure 8.3 *Not the result of a badly held camera, this view shows a Renault military vehicle on a bump rig. The 'bumps' are provided by means of electro-hydraulic rams.*

and has several bus stops, used in the simulation of urban driving conditions. In the centre of the circuit is a noise measurement area which is as far away as possible from the other test areas to minimise the effects of noise interference.

The Bedford proving ground is situated at Millbrook in Bedfordshire. Although it was originally a joint cars/commercial vehicles operation, under the banner of Vauxhall Motors, all connection with the car division has ceased since the restructuring of the whole European operation of General Motors. Accordingly, it is now referred to as the Bedford Proving Ground rather than the Vauxhall/Bedford Proving Ground, as was the case when it was first opened.

In total, the proving ground covers about 700 acres. Landscaping the site to build the track involved the movement of over 2.5 million tons of

Research and Development Facilities

earth. Apart from incorporating the required testing facilities, the building work necessitated planting over 200,000 trees and shrubs, partly for environmental reasons but also for security.

A bird's eye view of Millbrook shows that it is dominated by a circular high speed track, two miles in circumference, which consists of five traffic lanes plus a hard shoulder. This track is banked to the extent that the outer edge of what is effectively a saucer-like shape is some 15ft higher than the inner edge. The curvature of the track is such that each of the five lanes has its own safe 'hands off' speed, with that of the outer lane, for example, being as high as 160km/h (100mph). Bedford uses the high speed track to simulate high speed motorway running to assess durability, performance and safety.

A 137m (450ft) diameter concrete steering pad is situated within the circle of the high speed track and is used in the study of the steady-state steering characteristics of Bedford vehicles from Astra vans up to 44 tonne plus TMs. It enables the engineers to physically check computer predictions of cornering ability or the minimum tyre pressure needed to keep the tyre on the rim under emergency conditions.

The testing of tyre adhesion and vehicle handling on extremely slippery roads is carried out on a surface of Bridport pebbles covering a total area of 2,350m^2 (25,000 ft^2). Powerful jets of water sprayed on to

Figure 8.4 *Vehicle testing has not yet become totally independent of the weather, so many test tracks incorporate sections which can be wetted artificially.*

this surface give the correct low friction conditions for causing aquaplaning.

Of the two water troughs, again within the high speed circuit, one contains fresh water up to a depth of four feet and this is used for wading tests on the military vehicles in the Bedford range. The second trough contains a shallow saline solution (1 inch deep) simulating the effects of winter motoring on road spread with salt to clear ice and snow. To supplement the salt water in the trough, a battery of nozzles can direct salt spray over the sides of the test vehicle.

The mile-long straight at Millbrook is used for evaluating braking performance at maximum speeds on dry and wet surfaces as there is provision for wetting selected stretches. The approach road to the straight is slightly downhill with the result that, when cars were tested there, an initial speed of over 100mph was possible. This particular straight is also used by Bedford engineers to test acceleration and other performance factors as well as for studies on rolling resistance and transmission characteristics.

A number of special surface strips, laid out alongside the one mile braking straight, are used for studies of suspension behaviour, including the efficiency of damping and resistance to pitch. There are lengths of corrugated road with varying 'wave lengths' noise generating surfaces and another strip with random wave length and pitch.

Research and Development Facilities

Figure 8.5 *Off-road vehicles have to take a great deal of punishment in normal use and this is reflected in the manufacturer's testing techniques.*

As with most proving grounds, Millbrook features test hills where vehicles have to restart from rest – in both forward and reverse gears – on slopes with gradients of 20 per cent (1 in 5) and 25 per cent (1 in 4). In addition, there are 5.3km (3.3 miles) of hill routes where the gradients vary from 6.7 to 26 per cent (1 in 15 to 1 in 3.8).

A test of the truck's suspension and integrity of the cab structure is provided by the rough track which is an oval circuit some 870m (2,850ft) long surfaced with irregularly spaced $9in^2$ concrete blocks which protrude by either one or two inches. The trucks are driven over this surface, fully laden, at various speeds and for varying lengths of time.

Tippers and four wheel drive military-type vehicles are tested over a cross-country course. This is laid out on a hillside and includes a 1.5 mile section of rough 'road' with steep slopes and bends mainly on unmade surfaces. Thrown in for good measure are rocky sections, muddy areas and deep potholes, a severe section of two-inch sets and simulated ditch crossings.

The *pavé* section at Millbrook has an interesting history. The granite sets which make up the 1.5km (0.9 mile) long section came originally

127

We make heavy vehicles better in more than one way
Here's how

SAB AA1 AUTOMATIC SLACK ADJUSTER
Keeps brake adjustment 'spot on' automatically, reduces maintenance costs and downtime, improves safety. Fits S-cam brakes on buses, tractors, trailers and off-highway vehicles.

SAB TWIN CARTRIDGE AIR DRIER
Keeps air 'bone dry', improves systems life by reducing corrosion; filters air to remove contaminants. The complete 'air drier package'.

SAB ELECTRICALLY HEATED AUTOMATIC DRAIN VALVES
Operates with every air demand to remove moisture automatically from air reservoir. Helps protect against air system freeze ups.

PAUL DAHL AND KNORR BREMSE AIR BRAKE EQUIPMENT (concessionary products)
A full range of air brake equipment from compressors to load sensing valves for all types of systems.

SAB Automotive Co. Ltd.
Hilton Road, Aycliffe Industrial Estate, Newton Aycliffe, Co. Durham DL5 6SX. Telephone: 0325 310110. Telex: 587743.

---

from Edinburgh's Royal Mile and were ripped up when the city's trams were withdrawn. Over 3,500 tons of sets were used, each one laid by hand.

Bedford incorporated several of the Millbrook features into one 40 mile cycle designed to investigate the long-life potential of a prototype vehicle. A typical cycle would include a mile long lap of the *pavé*, several laps of the banked, high-speed track, two circuits of the hilly route and a pass through the salt water trough. Repeating this on a round-the-clock basis, fully laden, would result in some 65,000km (40,000 miles) being covered in six months, for a light van, and over 160,000km (100,000 miles) for a TM over a proportionately longer period.

One of the most dramatic of the outdoor test areas at Millbrook is where complete vehicles are crashed into a 75ton block of concrete or where vehicles are crashed into one another. This is one of a number of tests concerned with the safety of vehicle occupants and the ability of the body shell or cab to absorb impact while affording maximum protection. Complete vehicles are crashed into this concrete block at speeds of around 30mph. Up to eight high-speed cameras – including two in a pit directly below the crash point – record in detail the way the vehicle reacts during the impact. Metering equipment mounted on the test vehicle provides signals which are recorded on magnetic tape for later analysis in the laboratory.

Also used to assess occupants' safety is the indoor impact simulator at Millbrook, which Bedford claims was the first of its kind to be installed by any vehicle manufacturer anywhere in the world. The simulator works by inducing vehicle crashes 'in reverse'. Whereas most crashes on the road involve sudden forward deceleration, the simulator reproduces the same effect by producing sudden acceleration rearwards using about 250,000lb of thrust. Humanoid dummies are seated inside the body shell which is projected backwards from rest to speeds of up to 60mph in one tenth of a second – equivalent to around $65g$. High speed film, exposing 1,000 frames per second, records the behaviour of the dummies during the simulated impact. In addition, metering devices inside the dummies themselves supply data via multicore cables to magnetic tape recorders. The simulator provides information for such items as seat belt anchorages, steering wheel and steering column impact-absorbing properties as well as seat mountings.

In 1982 Perkins Engines of Peterborough became the first independent engine manufacturer to purchase a laser doppler anemometer (see Figure 8.7) to enable the research staff to measure the air motion generated inside the cylinder of a running engine. The equipment was originally developed at the Atomic Research Establishment at Harwell for measuring the flow of coolant in nuclear reactors. The system works by passing the beam of a laser through an optical device which splits the beam into two, at the same time changing the wavelength of one half relative to the other. When the two parts of the beam are focussed on a single point within the cylinder itself via a quartz window in the head, a pattern of light and dark interference fringes is created. Small particles are introduced which follow the air motion and scatter the light as they cross the bright fringes. The frequency of the reflected flashes is measured from which the air velocity can then be calculated. When sufficient points in the cylinder have been measured, a 'map' of the overall air movement can be produced. Perkins is using the laser doppler anemometer to validate a computer program which was developed from first principles to predict the air motion in an engine cylinder.

A great deal of Perkins' research activity is concentrated on the area of alternative fuels, as it is certain that diesel engines of the future will have to be capable of operation on a wider range of fuels. Already there is an increasing trend in some parts of the world towards a degradation in the quality of diesel fuel because of supply shortages which necessitate taking a wider 'cut' from each barrel of crude oil.

Fuel derived from coal will be a major energy source in the future for many countries, according to Perkins, and the Peterborough company

Figure 8.6 *Prototype testing is an expensive business. This is a Club of Four cab undergoing the impact test required by the Swedish safety regulations.*

Research and Development Facilities

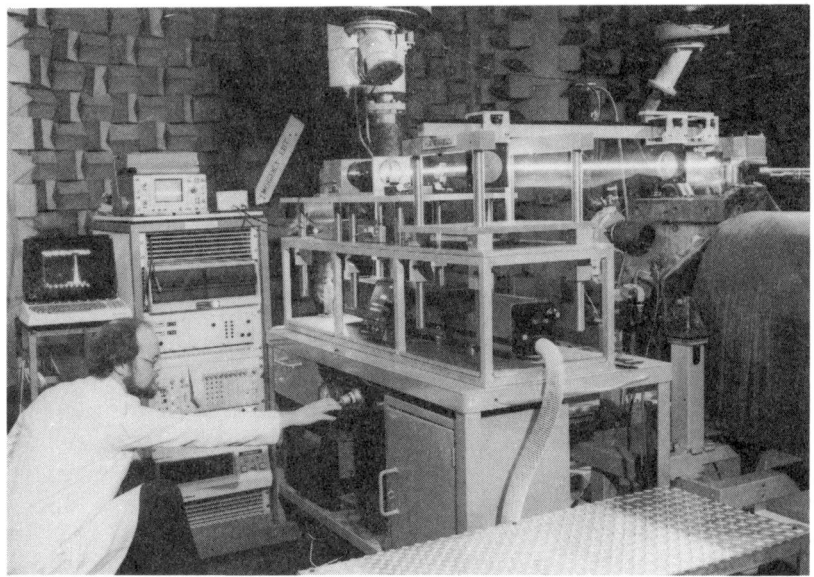

Figure 8.7 *This laser doppler anemometer allows research engineers at Perkins to measure the movement of air inside the cylinder while the engine is running.*

is devoting a lot of research into the problems of such a fuel produced either by synthesis or degradation. The results of early Perkins' tests indicate that both can give acceptable results when used in a diesel engine.

Alcohol fuels, eg ethanol and methanol, have very low cetane numbers (the cetane number of a fuel is a measure of its self-igniting ability) so, to make them suitable for use in compression ignition engines, an ignition improver must be added. However, conventional ignition improvers of the type added to low grade diesel fuel need to be added to alcohol in very large quantities and searching for alternative ignition improvers is another major part of Perkins' research.

Vegetable oils – such as sunflower, peanut, soya bean, corn and palm – have quite high cetane numbers and can be used satisfactorily in diesel engines, although the disadvantage associated with their use is that they have a tendency to coke-up the fuel injection nozzles. Chemical modification of these vegetable oils, aimed at reducing the molecular weight of one of the main components is currently being tried by Perkins in an effort to overcome this problem. The Perkins' view is that the diesel fuel of the future is likely to be a mixture of one of these alternative fuels and conventional crude-oil derived derv. In this way supplies of crude oil can be extended.

131

Operating on a multiplicity of fuels will not be without its problems, however, and one potential solution to this is likely to be the combining of the inherent advantages of the diesel engine and the spark ignition engine. The current research programme at Perkins has already involved the running of spark-assisted diesel engines using various ignition systems and a wide range of fuels.

Another aspect of alternative fuel usage, which Perkins is currently investigating, is electronic fuel injection equipment which will enable diesel engines to burn a wider range of fuel types.

The operating cycle of the engine is only part of the story. Current research programmes are investigating the type of materials from which the various engine components can be made. For example, Perkins is looking at the use of alternative forged steels for crankshafts and connecting rods – steels with a lower alloy content than those currently in use but with the same strength and resistance to wear and fatigue. In addition, the company is looking at simplifying the heat treatment process currently used for forged steel connecting rods, to provide the same mechanical properties but with a lower energy requirement.

Perkins is also looking at the use of spheroidal graphite cast iron for a number of components currently made from steel, including crankshafts and gears, while compacted graphite cast iron is being assessed for use on components subjected to severe thermal fatigue conditions.

Sophisticated testing and measuring techniques are no longer the sole prerogative of research and development departments; the more mundane world of production testing now makes considerable use of automation with the factories of Leyland and Perkins being good examples.

The Leyland automated engine test centre was opened in 1978 and cost £8million. It has a theoretical testing capacity of 750 engines per week and combines computerised control systems with the latest in mechanical handling equipment. The centre consists of five main areas: a pre-test area; a test cell area, comprising 64 cells arranged in eight banks of eight; a finishing-off area; a rectification area; and a despatch area, which also includes a spray painting booth.

Engines enter the centre on a carrier trolley designed to cope with any of the 200-plus likely engine variants. This trolley acts effectively as the engine's own test bed to minimise the time taken to connect up all the essential services. It travels on an automatic track which is embedded under the floor and consists of an outer loop with spur lines running to the cells. For final positioning of the engine in the test cell, the carrier is disconnected from the track and moved by means of an in-built hovercraft system using compressed air. Each engine is provided with a plastic identification disc which selects the appropriate test programme

Research and Development Facilities

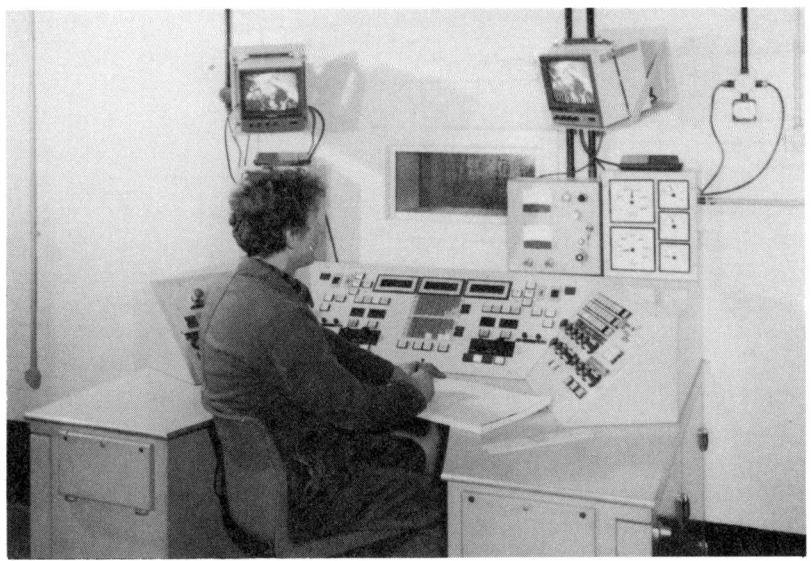

Figure 8.8 *Engine testing is becoming controlled more and more by computer. The operator acts as overseer to a number of test benches.*

for that particular engine specification from a computer data bank. The Leyland operator then selects 'automatic mode' on the control console and the computer takes over and runs the engine until normal oil and water temperatures are reached. A complete power curve test programme is then performed automatically, taking (and recording) readings of engine speed, torque, fuel delivery and exhaust emission. The computer then compares the results for the individual engine under test with a predetermined specification and only concludes the test when all the results are satisfactory. At this point, the engine is shut down with all the test data being transferred to the master computer. On completion of the test cell work, the engine is disconnected and the carrier is redirected to the finishing area.

Traditionally, prototype vehicles have been tested for performance in temperature extremes in places as far apart as Finland and Arizona, Canada and North Africa. Such tests include cold starting ability under sub-zero temperatures while, at the other end of the scale, the hot climates are used for evaluating the capability of the vehicle's cooling system. Such 'geographical' testing, however, does involve a considerable amount of time and expense – not to mention loss of secrecy – in transporting complete prototype vehicles about the globe, and because of this several of the larger manufacturers are using climatic wind

133

tunnels to evaluate vehicle behaviour under these diverse weather conditions. The use of the word 'weather' in this context is deliberate, as in addition to checking on the performance under extremes of temperature, the effects of humidity and solar radiation can also be evaluated.

The Fiat climatic wind tunnel complex at Orbassano near Turin is one example of the latest high technology being employed by a major manufacturer. Fiat uses two tunnels in fact; one to simulate tropical conditions, and one which is used for cold testing. The hot tunnel can generate temperatures of up to +50°C (122°F) while temperatures as low as -50°C (-58°F) can be achieved in the other tunnel.

Although the two tunnels are designated 'hot' and 'cold' they do have a considerable degree of versatility in that the hot tunnel can be used to produce temperatures as low as -10°C (14°F), while the cold tunnel can generate temperatures as high as +20°C (68°F), if required. Both tunnels share the same control room and, not surprisingly, a high degree of thermal insulation is required. The temperature variations are catered for by trichloroethylene/air heat exchangers located upstream of the test room.

In the hot chamber the humidity can be varied up to a maximum of 95 per cent using saturated steam sprayed in upstream of the fan which has ten variable pitch blades with an overall diameter of 5.0m (16ft 5in). Two rotational speeds of 75 and 150rpm can provide wind speeds of up to 160km/h (100mph). The temperature and humidity variations can be used to evaluate everything from coolant system capability, air conditioning and ventilation performance, cold starting capability, screen demisting and defrosting efficiency as well as monitoring the ventilation for the various loaded mechanical components.

To check on the durability of paintwork and rubber windscreen surrounds, an artificial 'sun' can be provided by means of a large bank of infrared lamps which can be varied in intensity from 500 to 1,200W/m^2.

Finally, dynamometer rollers in the climatic tunnels for commercial vehicle use can handle power outputs at the wheels of up to 450 horsepower.

Because of the sheer physical size of the company, Iveco (which includes the commercial vehicle division of Fiat) can boast of a number of establishments which can be included under the heading of research and development facilities. As well as the wind tunnel research centres at Orbassano, Iveco also has an engine research establishment at Arbon in Switzerland, a commercial vehicle experimental laboratory at Turin, a plant for building prototypes at Bolzano (adjacent to the Lancia Veicoli Speciali plant) as well as test tracks at La Mandria and Nardo in Italy and Markbronn in West Germany.

It will be interesting to see in the future just how many of the major manufacturers will continue to finance separate research into common problems, or whether the trend will be towards the establishment of independent research establishments which would act on a consultancy basis.

9. Road Testing Commercial Vehicles

Tim Blakemore

No matter how well designed, developed and manufactured a commercial vehicle might be, no fleet engineer is likely to consider it proven to his satisfaction until he has seen what it can do on the road. Transport engineers will listen with interest to manufacturers' claims of their new models 'extensive road trials' which have shown x or y improvement in fuel economy, performance and so on, but nobody is going to rush out and place a large order on that evidence alone. For most fleet engineers, practical men by and large, the ideal way of assessing a vehicle that is of interest to them would be to try one or more in their own fleets, and generally manufacturers are more than willing to provide demonstration vehicles for that purpose. Indeed, as competition among manufacturers has intensified recently, the length of demonstration period has tended to increase and it is now not uncommon for so-called 'seed vehicles' to be 'planted' into fleets for periods of as long as twelve months. But, of course, there is a limit to the number of demonstration vehicles that any manufacturer can have available at any one time, so not every operator will be able to try a particular model at the time he wants to. Conversely, there are many operators who feel that taking a demonstration vehicle puts them under an obligation to the manufacturer and so they are reluctant to do so unless they are fairly sure of eventually buying the vehicle anyway. So where does an operator, particularly a small operator (and that means the vast majority in the UK, for more than 80 per cent of vehicles registered here go into fleets of fewer than five lorries), turn for comparative data to help him reach his purchasing decision? One answer is Press road tests.

Currently, three leading trade journals in the UK conduct in-depth road tests of commercial vehicles. My description is confined to the methods used by Commercial Motor, for those are the ones with which I am familiar and indeed was partly responsible for establishing in their current form. Commercial Motor has five separate test routes (not including psv test routes), all on public roads, each route having been chosen to suit a different weight category of vehicle.

Without wishing to be pedantic I should make it clear at this stage that at Commercial Motor we prefer to describe the on-road part of each test as an 'operational trial' rather than a road test because that is

Transport Engineer's Handbook

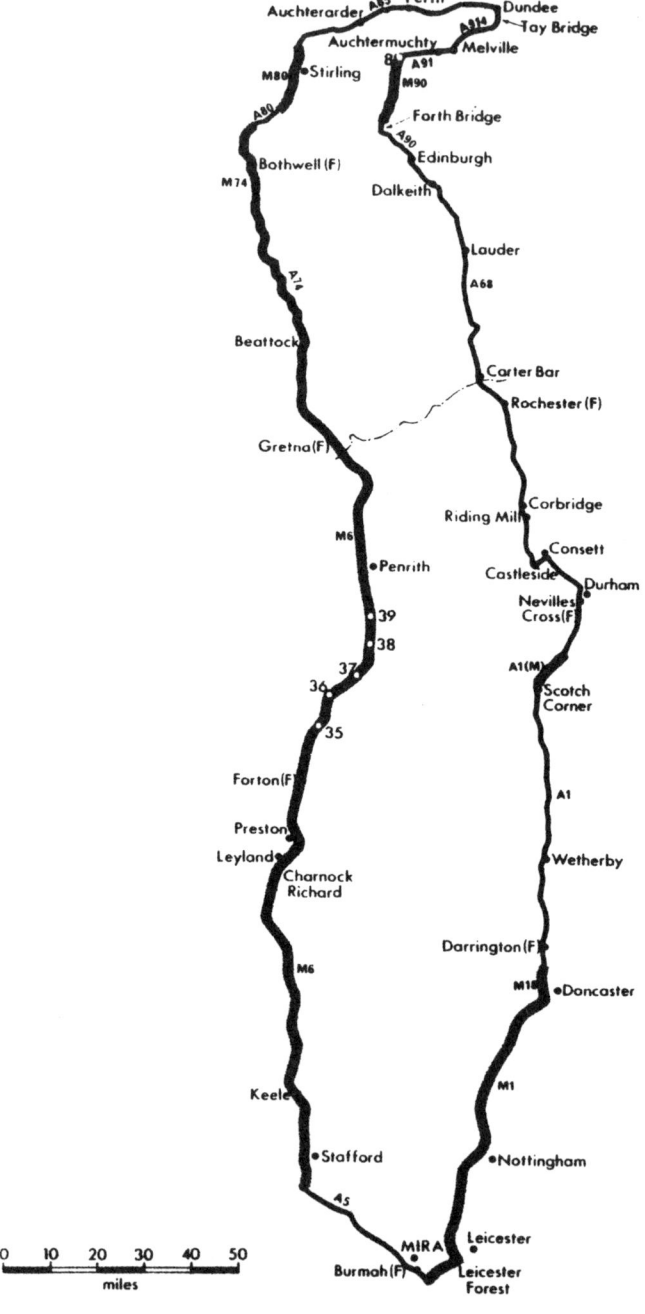

Figure 9.1 *Commercial Motor's 737 mile operational trial route for heavy vehicles, starts and finishes at a garage near the MIRA proving ground.*

Road Testing Commercial Vehicles

exactly what it is intended to be, and because we recognise that no matter how hard one strives for repeatable test conditions, certain factors are bound to vary from test to test when public roads are being used – hence 'trial' rather than 'test', though the latter word is often used in a more general sense.

The longest, and most established, of CM's routes is the so-called Scottish route which begins and ends near Hinckley, and goes as far north into Scotland as Dundee. This route is 1,195km (737 miles) long and is used to test haulage vehicles of 24 tonnes gross weight plus, whether rigid, articulated or drawbar; though almost all the vehicles tested over the past five years have been of the articulated type.

All vehicles tested by CM are in a fully laden condition, although light vans are additionally tested unladen, and for tractive units the loaded semi-trailer used will almost always be one of three kept for that purpose. Crane Fruehauf has provided us with a tandem axle box van semi-trailer since 1978 for testing tractive units at 32.5 tonnes gcw. When the maximum weight limit was increased to 38 tonnes, in 1982, York provided us with a tri-axle curtain sided semi-trailer for tests of 4×2 tractive units at the new limit, while Crane Fruehauf supplied another tandem axle model for tests of six wheeled tractive units at 38 tonnes.

The Scottish route was originally chosen to represent a typical route for a long distance haulier, which it still does, but from the operational trial results table, which is published with each test, an operator may also find it useful to pick out results from the section of the route which is of particular interest to him. Each results table shows the test vehicle's average speed over eight separate stages and average fuel consumption over seven separate stages, as well as day by day results for each of the three days. It is the overall average fuel consumption and speed which is probably of greatest interest to most operators and the unusual length of the route leads to these figures being very accurate indeed. Fuel consumption is measured by topping up the tank to a fixed point at each refuelling point, though most manufacturers nowadays fit flowmeters to their test vehicles and we do use these devices as a check against our results. Generally, modern flowmeters can be relied upon to give consistently accurate results and some manufacturers have satisfactorily solved the notorious fuel measuring problems associated with Cummins and Detroit Diesel engines, which have relatively hot fuel returning at a high flow rate to the tank. However, even though it is common now for a flowmeter result to be within 1.5 per cent of the 'tank top to tank top' figure, we never rely entirely on it and always use the total amount of fuel put into the tank as the basis for the final fuel consumption result. Where flowmeters can be particularly useful, once

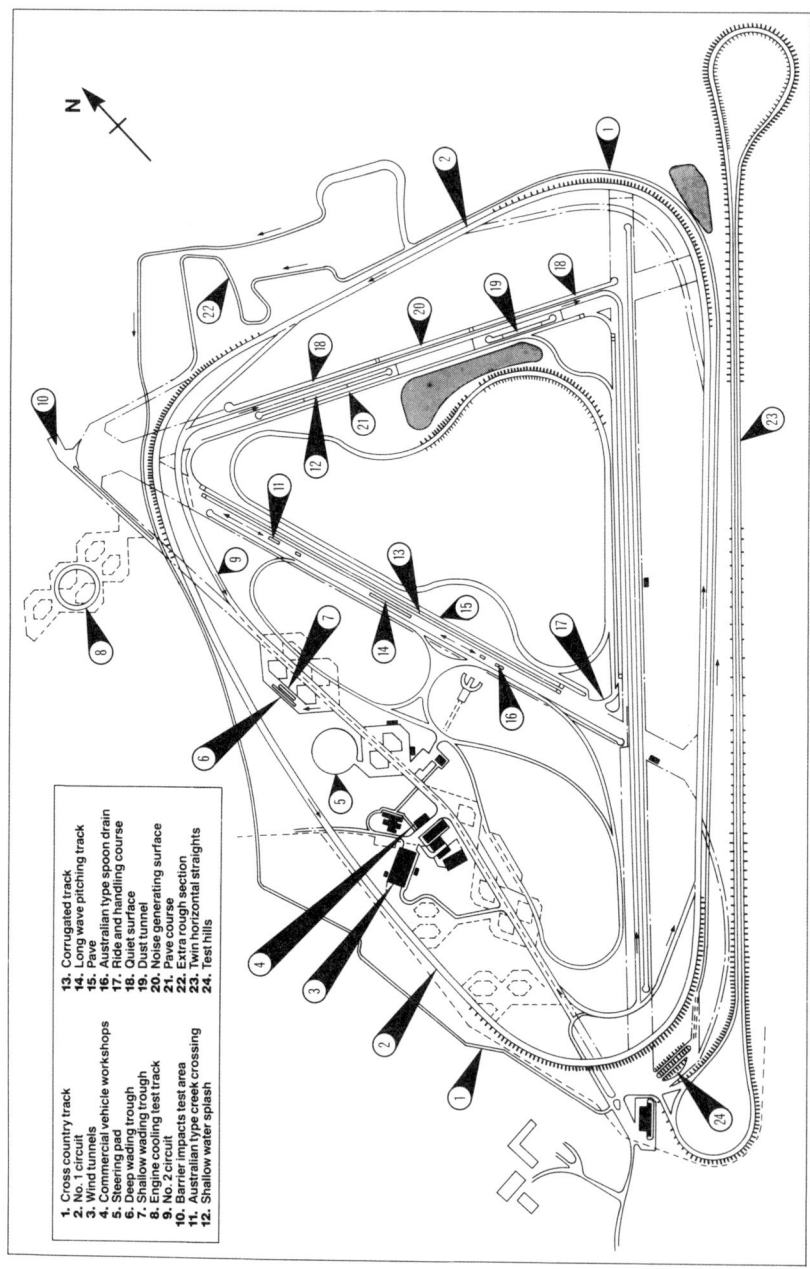

Figure 9.2 *Haulage vehicles with gvws from 7.5 tonnes to 24 tonnes are tested over this 210 mile route, starting and finishing at the Membury service area on the M4.*

proven to be performing properly, however, is in providing accurate intermediate fuel consumption figures, for, of course, refilling to a fixed point on the tank depends for its accuracy on the vehicle's attitude being constant and that cannot be guaranteed except when returning to the same spot on the original fuelling point.

Some engineers are surprised that we continue to use what, on the face of it, seems a fairly primitive method of checking fuel consumption, and certainly a bent length of welding rod for dipping a fuel tank could hardly be described as high technology, but practical experience has shown that with care very accurate results can be obtained in this way without resorting to weighing the fuel used or using any more sophisticated means of fuel measurement.

One of the biggest variables on any road test is weather and it goes without saying that it can have a considerable effect on the results, especially fuel consumption. Some manufacturers have estimated, for example, that the effect of rain and wet roads on the fuel consumption of a 38 tonne gcw tractive unit and tri-axle semi-trailer can be as much as 10 per cent. While the effect of varying wind strengths and directions on a vehicle, which itself is not following a straight path, is much more difficult to quantify, it is equally clear that a strong headwind from the general direction of travel of the vehicle will have some adverse effect. Rather than attempt to measure the effect of weather we simply accept that it is a variable beyond our control and let the reader form his own judgement on what part it played, from the brief summary of weather conditions we publish for each of the three test days. The meteorological data including wind speed, direction and ambient temperature comes from a print-out from the microcomputer at the Motor Industry Research Association's proving ground.

One major variable over which we *do* have control is the driver, who is always a CM staff member. The general rule on tests over the Scottish route, and indeed every CM route, is that we follow the letter of the law so far as speed limits are concerned so, for example, heavy vehicles are driven at no more than 40mph on A class roads and at 60mph on motorways. Where manufacturers make recommendations regarding certain engine speeds, for good fuel economy for instance, we will follow them, provided that does not mean that the vehicle has to be driven in an abnormal manner. We now publish a segment of the first day's tachograph chart from each heavy vehicle test so that readers may compare different vehicles' speed traces over a given section of road, in this case the northern part of the M6 which includes the long climb over Shap.

The time taken by a heavy vehicle to climb some of the steeper hills can reveal a lot about its 'on road' performance and is the kind of information an engineer cannot find in manufacturers' data sheets. Hill

Figure 9.3 *CM's light vehicle test route was changed at the beginning of 1983 to reflect the generally increased use of motorways by vehicles of this kind.*

climb times are recorded for five different gradients on the Scottish route, varying from the long, steady 1.7 per cent gradient of the M18 motorway to the very steep Blackhill climb through Consett, which has an average gradient of 7.2 per cent. The test vehicle's hill climb times are published in a table with those of three or four rival vehicles in the same class, so that readers may see at a glance how it compares. Whether the test vehicle is a car derived van or a 38 tonne vehicle the first part of its test will be conducted off public roads at the MIRA proving ground near Nuneaton. Axle weights, kerb weight and gross weight are checked, and the latter adjusted if necessary. It is not unknown for manufacturers to make rather surprising errors in loading their vehicles to their maximum legal gvws. (One manufacturer provided a 16 tonner, that is with a gvw of 16.26 tonnes (16 tons), loaded to only 16 tonnes.) Then the speedometers or tachographs have to be checked for their degree of inaccuracy, since most of the subsequent tests cannot be properly conducted without knowing the difference between indicated and actual speed. This check is carried out on twin horizontal measured mile straights, using a stopwatch to measure the time taken to cover the distance at various speeds. These twin straights are then used for the acceleration tests. The published acceleration times are always the mean of several runs, to allow for the effect of wind strength. With multi-ratio gearboxes, several runs are often needed to establish what is the best gearchange sequence.

If the vehicle is an articulated combination the overall length is measured to ensure that the maximum legal limit is not being exceeded. Now that the limit is 15.5m it is much easier for manufacturers to

Road Testing Commercial Vehicles

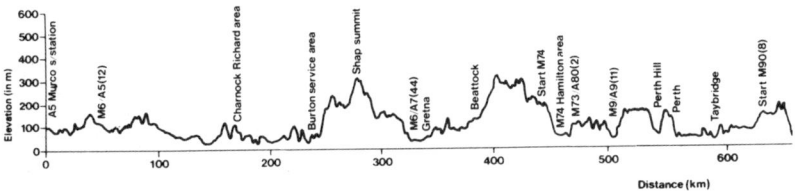

Figure 9.4 *MIRA's extensive proving ground facilities include twin, horizontal one mile straights which are used to check speedometer or tachograph accuracy and for acceleration tests. The 4.5km (2.8mile) high speed No. 1 circuit has banked corners which allow vehicles to be driven at their maximum speeds.*

comply with it, but in the days of the 15m limit some last minute adjustment of sliding fifth wheel couplings was often necessary. The fifth wheel coupling height is also measured, for now that we have a 4.2m height limit that, too, can be a critical dimension for many operators.

MIRA's extensive proving ground facilities include a banked high speed circuit, 4.5km (2.8 miles) long, which allows vehicles to be driven at speeds in excess of 100mph. Commercial Motor has never been presented with a commercial vehicle capable of quite that speed, but some light vans can attain speeds in excess of 90mph. Whatever the vehicle, we check its actual maximum speed on this circuit, which is also used when in-cab noise levels are measured. These are checked by holding our noise-level meter at the driver's left ear and driving around the circuit at a number of steady speeds. The objective is to allow readers to compare the in-cab noise levels they can expect from different vehicles under normal operation.

MIRA's ride and handling course, which has a wide variety of road surfaces, including a simulated rail crossing, pot holes, ruts, a 'washboard' surface, and a number of different cambers and bends, is used to help assess a rigid vehicle's ride and handling characteristics, though behaviour on our set routes can also reveal a lot about how well-suited to its task is a vehicle's steering and suspension.

Foot brake performance is checked in two ways, dynamically on MIRA's 'No. 2' 3.7km (2.3 mile) circuit with full pressure stops from 20, 30 and 40mph, and statically on a rolling road brake tester. This latter test was recently introduced into our testing programme to give a better, more reliable and consistent basis for comparison of braking figures.

A major advantage of the rolling road test is that it quickly shows up any lateral braking imbalance or any excessive brake drag, and with an articulated combination shows how braking effort is divided between tractive unit and trailer.

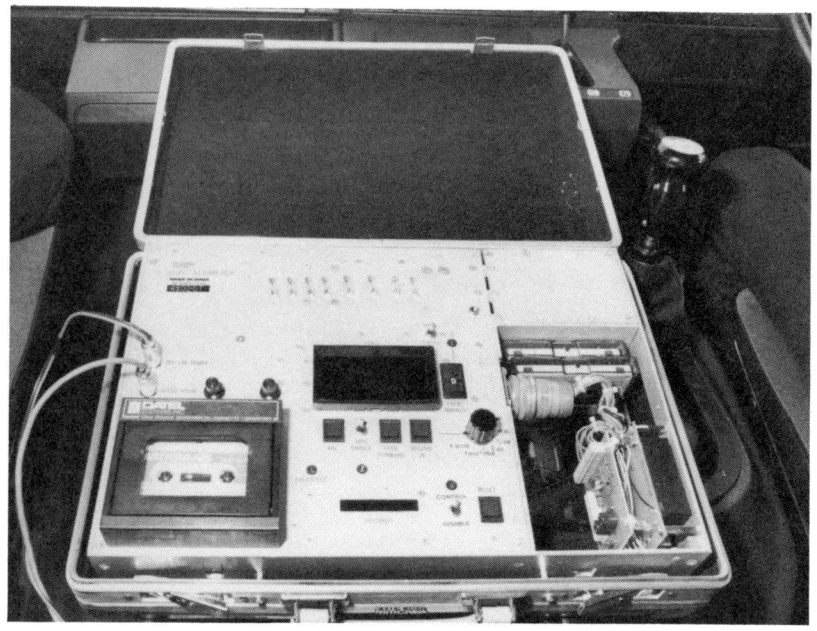

Figure 9.5 *Park brake performance and restart gradeability are checked on test gradients. The actual gradeability usually agrees with the theoretical gradeability but sometimes badly adjusted controls can cause difficulty.*

MIRA has four test hills with gradients of 1 in 6, 1 in 5, 1 in 4 and 1 in 3. These are used to test each vehicle's park brake performance and its restart gradeability. It is interesting to note that some vehicles fare badly on the test hills not because their theoretical gradeability is poor, but simply because they have insensitive or difficult to control clutches, throttles and/or park brakes.

Most manufacturers will have spent some time preparing a Press test vehicle and so will know what to expect from our formal test procedures. However, one of these procedures produces figures which rarely agree with those to be found in brochures or specifications. This is the measurement of turning circles. The first problem is that manufacturers quote one dimension only for a turning circle when in practice there is almost always a difference between left and right locks. We measure both by chalking the circle described by the outer edge of the tyre as the vehicle is driven slowly on full lock. That gives us the kerb-to-kerb turning circles. The wall-to-wall turning circle comes from measuring the arc swept by the vehicle's front corners. The on road part of each road test, the operational trial, does much more than provide a set of fuel consumption and average speed figures, it gives the tester his

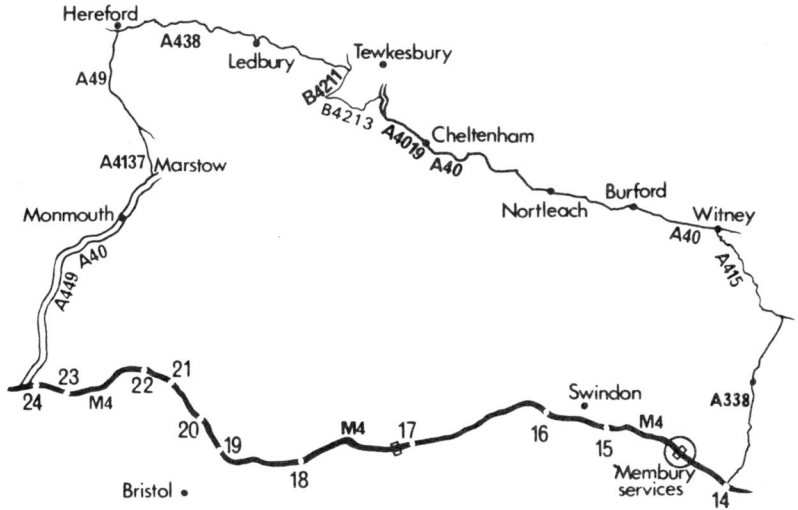

Figure 9.6 *Several vehicle manufacturers have begun to use computers to simulate road tests. This profile of CM's Scottish route was drawn by Scania's STRASS (Scania Transport Simulation System) computer.*

opportunity to form an opinion on the vehicle, which will be the basis for the test report.

In the layout of road tests in Commercial Motor we now deliberately separate opinion from the pure facts and figures, recognising that certain readers will be more interested in one than the other. A five page heavy vehicle road test, for example, has three pages of words and pictures describing what the tester considered the most important features of the vehicle and two pages of 'pure' data. One of these pages is devoted to what might be described as fixed data, that is dimensions, specification, weights and so on, as well as some information about the vehicle manufacturer's service support and some typical parts prices and workshop times. The last page has all the data gathered from the test and operational trial and some histograms which show how the vehicle compares in critical areas with its major rivals.

Several manufacturers are currently striving to take the subjectivity out of many areas of their own vehicle and component testing, including road tests, and not surprisingly computers feature strongly in their work. Scania, for example, has a system called STRASS (Scania Transport Simulation System), Daimler-Benz has its TRASCO (Traffic Simulation by Computer) and Cummins has its VMS (Vehicle Mission Simulator). (Scania's engineers are now so satisfied with the accuracy of their computer program that they hardly need to road test vehicles at all!)

Transport Engineer's Handbook

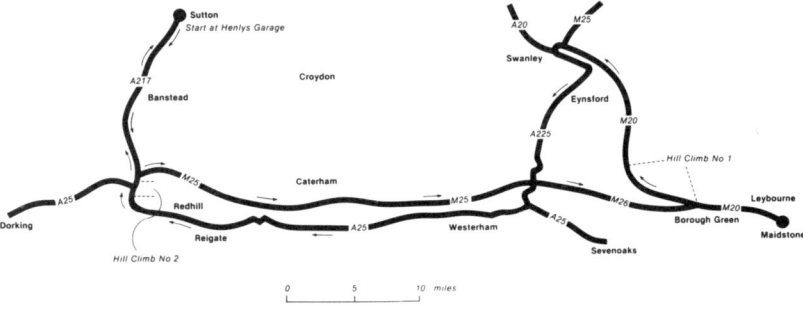

Figure 9.7 *The data for Figure 9.4 was collected using the very compact 'Road Altimeter' shown in the photograph.*

A special feature of the Scania 'Road Altimeter', which is used to collect data from any selected route, is its compactness – it fits into an alloy attaché case and only two connections have to be made to a 12V power source and to the tachograph drive (see Figure 9.5). Each revolution of the tachograph drive sends pulses to a magnetic tape, while simultaneously, a remarkably sensitive altimeter senses changes in elevation (a change in height of as little as 0.5m (20in) can be detected) – this information is then fed on to the same tape and an accurate profile of the route can thus be plotted (as shown in Figure 9.7). To discover how any of its vehicles will perform over a recorded route, Scania simply runs a tape containing data on all the vehicle characteristics such as rolling and air resistance, engine power, torque and specific fuel consumption, gear efficiencies, rear axle ratio and so on, through its computer and the results are produced with much less time and effort than a 'real' road test takes.

A significant advantage of this system is that the effect of changing any of the vehicle characteristics can be very quickly discovered. The results are so detailed that Scania's engineers can predict, for example, how many gearchanges need to be made on a given gradient, where the optimum gearchange points are on that gradient, what the effect would be of changing down earlier or later, and, of course, how much fuel will be consumed.

All this inevitably begs the question 'Is there now any need for independent human road testers?' Perhaps not, particularly if computers could be relied upon not to write: 'the gear lever fell readily to hand'.

10. The United Kingdom Bus and Coach Market

John Taylor, Editor, *Coaching Journal* and *Bus Review*

There is something of an ambivalent air about the British psv market. On the one hand, the sale of buses – both single and double-deckers – for stage services continues to stagnate. On the other hand, while the sales of coaches are still down compared to 1982, the decline is not all that great and the expenditure is much greater as rising specification requirements are reflected in the unit cost. For the record, the total number of bus and coach registrations in 1980 was 5,792. In 1981, the total was 4,441 and in 1982 it was 3,766. Taking the 1983 figures to the end of September, the total number of psvs registered was 3,048 against 3,144 for the same period in 1982. Extrapolating these figures, one could project that the final total for 1983 would be 3,651. Thus, taking 1980 as 100 per cent, the market fell to 63 per cent; perhaps not as bad as the commercial vehicle market, but still a poor outlook for our domestic factories, especially in view of the increased competition from Europe.

When it comes to the European and Scandinavian market share of British psv sales, this has increased from just under 20 per cent in the first nine months of 1982 to almost 24 per cent for the corresponding period of 1983, so there is little cause for complacency however one might look at the situation. The problem is compounded by the decline in psv exports because of the strength of Sterling, which has also encouraged European chassis makers and bodybuilders to turn their attention to the UK market. Indeed, with some coachbuilders, the flourishing UK market has been their salvation with depressed sales at home.

As far as the traditional stage carriage bus market is concerned, the phasing out of the New Bus Grant has been another brake on sales, as is the restriction on local government expenditure. Single-decker sales have declined to almost insignificant proportions and double-deckers, once the mainstay of UK factories, are also in poor fettle. Leyland, who have traditionally dominated this field, have had to 'cut their cloth' accordingly. The AEC works at Southall and its associated Park Royal bodyworks had already closed and come under the hammer (literally) for redevelopment. The Bristol works have also closed and production of the Olympian chassis has been transferred to the Workington plant

Figure 10.1 *Production of the Leyland Olympian double-deck bus chassis was moved in 1983 from Bristol to Workington. This is the first of the model to come from the new location. Bodywork for the Olympian comes from Eastern Coachworks at Lowestoft.*

which was originally designed for the Leyland National single-decker. This has been joined, in recent years, by the Titan integral double-decker, also transferred from Park Royal. Now, production of the advanced Titan is destined to end in 1984 because of lack of demand, and the Olympian will in the future be the main Leyland double-decker.

One reason for the demise of the Titan is London Transport's lack of further interest, caused by escalating costs. London Transport bus policy is now in the melting pot but future demand will almost certainly be satisfied by standard products. The coming year will see trials of various types of buses, including the Olympian, and the MkII MCW Metrobus – LT already has a large number of MKIs in service – and the Volvo Ailsa. The capital's disenchantment with the Fleetline, of which it had several thousands, has further complicated the market. These have now been phased out and sold off in large numbers, some being scrapped years before their time, but most finding new lives with British and overseas operators. National Bus companies, the Passenger Transport Executives, municipal undertakings and independent bus operators have all taken the much maligned Fleetline and in most cases have found it eminently satisfactory. The same is now happening with the MCW Metropolitan, which London is selling and which others are now snapping up. Quality secondhand buses – and once overhauled and modified to their new owners' requirements, these *are* quality vehicles – mean fewer potential sales for companies selling new double-deckers, and so the decline continues.

Despite this, there is still interest in the market and Dennis in particular has gained a healthy share of what cake is left, while Volvo's Scottish-built Ailsa finds a steady demand. Scania has been trying to tempt buyers with its semi-integral double-decker chassis and are selling a few, but the most interesting innovatory design has again come from Volvo. Again, because the Irvine factory originated the Ailsa when the market settled on the rear-engined design. The Ailsa used a multi-channel welded peripheral frame with a front-mounted engine. This was made possible bcause of the slim cross-section of the Swedish-built power unit, though even then the driving compartment was not over-generous.

Late in 1982, Volvo launched its Citybus, a clever development using the same frame principles as the Ailsa but with the slim engine in horizontal form mounted amidships, as on the B10M coach chassis. This gave the front end space of a rear-engined bus, with an acceptably low floor height and made full use of the rear space taken up by the engine compartment on other designs. Operation of the prototype in Glasgow, with Strathclyde Passenger Transport Executive, has proved promising enough for other transport executives to order the Citybus in small numbers for trial in their own regions. Certainly if it does work well in practice it could change the whole approach to double-decker buses, as did the rear-engined machine, and bring it into line with underfloor-engined single-decker buses and coaches, which still take up the bulk of British sales.

Turning to the coach market, the situation of a year ago has continued without radical change, though the future holds some interesting promises. The market still continues to be dominated by heavyweight vehicles. Ford and Bedford continue to lose out on sales and between them their share for the first nine months of 1983 had fallen to 13.25 per cent of the whole (including buses) as against 16.22 per cent during the same period of 1982. Quite what the coming year holds for either company is open to conjecture. Bedford has high hopes for its new 12 metre chassis due to be launched in the autumn of 1984, but whether the majority of customers will have by then made up their minds to go elsewhere is a moot point.

The big question mark must hang over Ford. Production of the psv chassis is only a tiny part of the activity of the Langley truck plant near Slough – indeed, it is said that the entire year's demand can be bulked through in one or two days. The future seems to hinge not on activities at Slough or decisions at Brentwood, but on developments in a factory in Telford New Town.

Figure 10.2 *This special low entrance bus developed by Ralphs Coaches for Heathrow Airport work on a Quest 80 chassis holds promise for the future.*

Quest 80, a company formed to design vehicles for the South African market, has been pursuing energetically the UK coach market. After initial development work with service buses, the company gained an order for 20 high floor chassis from Excelsior Holidays, the major tour operator based in Bournemouth. Other business seems likely to follow and a notable development was the chassis engineered jointly with Ralphs Coaches to provide a low entrance vehicle particularly suited to their airport passenger transfer work at Heathrow Airport, though the wider implications are readily apparent.

Where the Excelsior contract fits into the pattern of things is that the company pioneered the Ford psv chassis from its introduction over 20 years ago and has been responsible for much of the in-service development work. Excelsior has run an all-Ford fleet until recently, since when it has taken DAFs for the heavier duties which the work involved. These rear-engined chassis have been very successful and now the introduction of the Quest into the fleet will add a rear-engined vehicle with Ford components.

Quest chassis vary according to requirements but are based on the Ford Cargo vertical engine, mounted for and aft at the rear of the chassis, offset and driving the rear axle by means of a duplex chain. The Excelsior contract calls for the 2728T engine, modified by Sabre and

The United Kingdom Bus and Coach Market

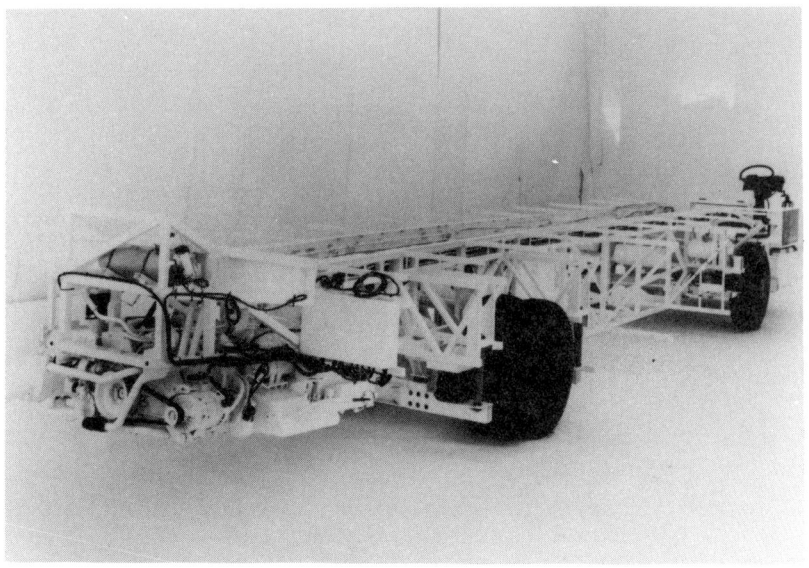

Figure 10.3 *Mechanical components on an integral coach are still mounted on a sub-frame as on this Leyland Royal Tiger. It is then integrated into the body structure.*

Quest, as a charge-cooled turbocharged unit producing 165kW (220bhp) at 2500rpm. Full air suspension is a feature of this model, designated the VM, though leaf springs are available if required. Whilst Ford is saying little about its direct involvement in the psv market, it seems more than a distinct possibility that once the Quest is proved in service, it might drop their own range and endorse the Telford product, a move that would give Ford an updated range without the problems of actual production.

Where Bedford is concerned, the YNT has now settled down after teething troubles and since becoming available with the ZF gearbox obviously has plenty to commend it until the new design comes along.

Despite the almost lemming-like rush towards the top-quality, heavy-duty coach, there is still a very worthwhile market for the unpretentious middleweight, simple in design and easy to maintain, ideal for day and half-day excursions, darts club outings and school bus work – in short, the bread and butter work of many a coach operator.

The quality market is by no means single-minded in its choice of product. Underfloor mounted engines still predominate but rear-mounted units are gaining ground. High-floor 12 metre designs are very popular, offering large luggage capacity for shuttle services to the

Figure 10.4 *This Plaxton Paramount body in 3.2m height form on a Leyland Leopard chassis is typical of the modern British coach.*

Mediterranean and ski holidays that are a feature of so many operators' work. Integral designs top the market, though the separate chassis still takes the lion's share, while in between there is the chassis split into front and rear subframes for incorporation into an integral style body. Certainly the 12 metre coach is very much in vogue and 3.5m (11ft 6in) height is the current norm for many, others sticking to 3.2m (10ft 6in). The articulated coach never came to anything because of the French authorities banning its use and so destroying its economics. What has come in with a bang is the double-decker and the neo-double-decker.

The full double-decker coach offers a great deal for the shuttles mentioned above, and for express services in Britain. Actually, the latter are by no means new as Ribble/Standerwick used them on motorways back in the early 1960s. Today's full double-decker is a very different kettle of fish. Carried on three axles and 12 metres long, it seats around 74 passengers, mostly on the top deck. The lower floor is given over to a large rear end baggage room, toilet, driver's sleeping berth and, amidships, a lounge with tables. Such vehicles are used by Trathens on its Plymouth and Exeter expresses from London, and now National Express and other NBC companies are introducing special versions of the Leyland Olympian on such services. The MCW Metroliner, which made its debut at the 1982 Motor Show in Birmingham, has been in continual service with the Scottish Bus Group on its services between London and Scotland and has proved to be a highly successful and very reliable coach for what was essentially a prototype. As a result, MCW has received orders for a further six for the Scottish Group and 39 for

The United Kingdom Bus and Coach Market

Figure 10.5 *The neo-double-decker, with its high floor, large baggage compartment and rear downstairs lounge is becoming quite popular for long-distance work. This is a Jonchkheere body on a Volvo B10M chassis.*

National Bus.

For a country that has been so clearly identified with the double-decker it is ironic that the new generation of double-deckers come from the Continent. The Trathens vehicles are built in Germany by Neoplan – who also supplied the two luxury sightseeing versions to Harrods store – and Van Hool and Kässbohrer also sell on the UK market. Plaxtons, the major British coach builder, is also producing a double-decker using a Neoplan underframe. The overall market will not be great because the vehicles will have to last to recoup their investment. On the other hand, one coach will do the work of two 49 seaters on a shuttle and save on drivers and operating costs. Coaches used on this sort of service regularly cover 4,000 miles *a week* by doing a couple of shuttles to Spain and back, and some operators are claiming a full 200,000 miles a year from some vehicles. Such intense utilisation would have been impossible only a few years ago and it is a tribute to the reliability of the modern coach that it can give such service. Of course, spending anything up to £140,000 on a double-decker coach does demand that sort of use to recoup the original investment.

The conventional separate chassis still prevails in 11 and 12 metre forms and Leyland dominates the market with the Tiger. However, competition is rife and Volvo claims second place and very strong allegiance from the independent operators with its B10M design. In a sense, it has inherited the mantle of the AEC Reliance as the alternative

Figure 10.6 *A super deluxe MCW Metroliner coach now in service with the Scottish Bus Group. The Express version has additional seats and no centre doorway.*

to the Leyland and, although now history, hindsight indicates that had the Southall company been allowed to pursue its own destiny, even under the BL banner, the overall picture might have been healthier.

Catching up rapidly, and currently in third place, is the Dutch company DAF which has been getting some quite big fleet contracts from Excelsior, Shearings and other major operators. The Eindhoven company hedges its bets nicely by offering underfloor-engined and rear-engined chassis. Leyland does the same with the integral Royal Tiger which is now in production and coming into operational service. While the Doyen body, built by Leyland subsidiary Charles H Roe at Leeds, is designed for the Royal Tiger, the framing can be had for bodying by other builders and Plaxtons has completed one for their Paramount 3500 design.

Where British manufacturers are concerned, the dark horse is MCW. This Birmingham-based company, long famed for its service buses and railway rolling stock, has been feeling the draught with the contraction of its traditional markets.

It was this that prompted its current foray into the luxury coach field, having had several abortive attempts in earlier years. The new generation of six wheel double-deckers and the integral rear-engined single-deckers, all with Cummins engines, are premium quality vehicles and the manufacturer has taken the right steps to market them by appointing a recognised coach dealer who will be in a position to cope with trade-ins, as with any other make. How this present attack on the coach market, better planned than any before, will fare is still open to conjecture but it certainly deserves a share of success.

DAF is not the only representative of Dutch manufacturing on the UK market. Bova is a company that builds integral coaches using DAF mechanical components and it has been remarkably successful in Britain over the past three years, again with the backing of a major dealer who has now, in fact, actually bought a major shareholding in the Dutch company. Bova's Europa coach has been bought by many fleets including National Travel, Wallace Arnold and other big names. Now the company is launching its Futura in Britain, a very different approach to the plain, boxy lines of the Europa. The Futura is a luxurious 3.2 and 3.5m coach with a very rounded frontal treatment aimed at giving good fuel economy. Its looks were something of a shock at first but there are signs that other makers will soon follow suit. Meanwhile, Moseley, the dealers for Bova in Britain, have been working with Duple Coachbuilders to put a 3.2m version of the Caribbean 3.5m body on a Europa underframe. Called the Calypso, this holds considerable promise for companies that have been pleased with the Europa but wish to ring the changes on body styling.

When it comes to coachbuilding, Holland and Belgium have become dependent on the British market in recent years. Jonckheere sells very well in Britain on Volvo, DAD and Mercedes-Benz chassis. Van Hool is also finding integral and separate chassis designs in strong demand. The smaller bodybuilders such as LAG, Berkhof and Smit have all found Britain rewarding, due in no small part to the current exchange rate which favours imports and inhibits British companies' export activities. How some of these companies would fare if the exchange rate changed dramatically in Britain's favour is open to question.

Spain and Portugal have done steady business in Britain in recent years though over the past year or two this has flagged noticeably as Dutch companies have moved in. However, Caetano of Portugal has a new model and it is possible that Ayats and Irizar from Spain may have another attempt after earlier and unsuccessful ventures on the British market.

The French have kept clear of the British coach market, though a question mark hangs over the new Renault design which the truck and bus division is considering exporting to Britain. The Italians have also steered clear and Fiat has confirmed that it has no plans for a full size

Figure 10.7 *Fiat sells only small coaches on the UK market like this Portuguese-bodied 18-seat Beja.*

coach in the UK unless conditions change sufficiently to make it an attractive proposition. Meanwhile, the smaller chassis are being bodied as coaches by Robin Hood Coachbuilders of Southampton, who have the attractive Krypton design, and Caetano who export the 18 seat Beja. The only Italian coachbuilder to sell in Britain is Padane, whose handsome Z design has achieved modest success.

Then there is Germany. The connotation of quality is maintained throughout when one deals with names like Mercedes-Benz, Kässbohrer and MAN. Kässbohrer has a Rolls-Royce-like status in coaching and the very high technical specification allied to quality of construction and equipment makes this a prestige product, a point borne out by its many successes in international coach rallies.

MAN has found mixed fortunes with the SR280 family. There is certainly nothing wrong with the vehicle and the quality and high standard of equipment are there. The trouble was that the UK importers insisted on dealing directly with buyers, which was a mistake. Now the company is fully under the VAG banner and has appointed a dealer network that can cope with the trade-ins which are so important in clinching a deal.

Mercedes-Benz was among the first to enter the UK market with trucks and coaches, but for many years concentrated on the former and let the latter die. Recently, Roeselare Sales has been importing Mercedes chassis with Jonckheere bodywork and now Britain's biggest

The United Kingdom Bus and Coach Market

Figure 10.8 *Air-smoothed shapes for fuel economy are becoming a feature of modern coach design, as exemplified by this Duple Laser 53-seater on a Bedford YNT 11m chassis.*

coach dealer, W S Yeates of Loughborough, has taken on the exclusive franchise to sell the integral 0 303 coach in Britain. This has been a long time in coming on to the British market (Wahl, the London-based, German-owned company has a number imported direct), but is likely to upset the present fine balance and create a healthy demand.

Another German company, already mentioned, is Neoplan, whose distinctively styled single- and double-decker coaches are already becoming familiar with British fleets. Probably the only major coach-builder left to come to Britain is Drögmöller, a very old established coachbuilder based in Heilbronn who builds anything up to three axle double-deckers. Whether it will make the move is one of those points of conjecture that makes the coach market so fascinating at present.

Although Volvo has tended to dominate the thinking as far as Scandinavia is concerned, Scania is now making strenuous efforts to sell the K112 coach chassis in Britain. This design, like its double-decker bus counterpart, is rear-engined and is manufactured with a very short wheelbase, the idea being that the coachbuilder cuts it in half and integrates both ends into a complete integral body.

Back home, mention must also be made of Dennis. This name is one of the oldest in commercial vehicle manufacture and since being taken over by the Hestair group has shown strong signs of revival. Apart from its headway in the bus field, it is also moving on to coaches, and several are now in service with National on the Rapide routes.

Our two major luxury coachbuilders have had mixed fortunes. Plaxtons of Scarborough finally shook off their old and very long lived designs last year and came up with the Paramount (see Figure 10.4). This is a continental style body in 3.2 and 3.5m heights with several permutations of equipment to suit differing market needs. It lends itself to separate chassis, from Ford to Volvo, and can be integrated on Leyland Royal Tiger and Scania componentry. Certainly, it has proved very successful in its first season and the addition of the double-decker could well be augmented by further variations.

On the other side of the North of England, in Blackpool, Duple has been going through a traumatic period. The company had invested heavily in retooling and at the 1983 Motor Show came out with two completely new designs. The high floor model was the Caribbean, a good looking modern touring coach in the continental manner. For those wanting a more general purpose vehicle of standard height, there was the Laser, a notably rounded shape that promised good fuel returns. At the same time, the company continued to build its Dominant coach range that had been in production throughout the 1970s and was only finally phased out late in 1983. Despite considerable interest, orders were slow to come in for the new coaches and a major restructuring was thought necessary. This came about with the acquisition of the company by the Hestair group, which also owns Dennis. Production is still gearing up, but the outlook is certainly rather brighter, though 1984 could well be the year that really decides the future of the old established company.

In many ways, the current luxury coach scene is rather like it was thirty years ago, with many chassis and body makers competing for orders. Like then, a rationalisation looks likely before too long.

11. Keeping Wheels Secure

John Dickson-Simpson

There are few operators of heavy trucks and buses free from occasional problems of wheel-stud breakage, wheel looseness or wheel-cracking – sometimes even wheel loss. Some operators suffer more than others and it is not unusual for six per cent of hubs to be affected in some way.

The manufacturers blame poor maintenance. The operators blame unscientific design, conflicting advice and demand for maintenance that is unrealistically meticulous and frequent. Many manufacturers advise checking wheel nuts weekly whilst operators claim that there is something wrong with anything that has to be checked so often.

Concern about wheel security is understandable. It is somewhat fundamental to the vehicle's mobility that the wheels stay on. It is also no little matter of public safety. Although the incidence of wheel loss is statistically tiny, it would be preferable if it were much smaller. The police are taking a close interest since there are twice as many reports of detached wheels today as there were ten years ago.

Wheels are sometimes lost because of hub-bearing nuts coming off or because of cracks of the wheels themselves. Mostly, though, the cause is broken studs. And broken studs are frequently a result of nut slackness – not always serious slackness.

There are two broad types of fastening for disc wheels. In one, conically-ended nuts locate the wheel on chamfered stud-holes, as well as clamping the wheel in place. Variations on the theme use plain nuts with tapered washers or spherical-profile washers. In the other broad type the nuts have plain washers and they simply clamp the wheel in place; it is loose-fit, located in the centre by a spigot on the hub.

With both arrangements, wheel security depends greatly on friction between the clamped surfaces – entirely so in the case of spigoted fixings. If there is insufficient friction to resist the dynamic forces on wheels or the rotational forces from brake and drive torque there becomes risk of stud breakage.

With coned fixings, forces are diverted through the nut, producing cyclical bending of the studs and fatigue failure. With spigoted fixings, the wheel rocks backwards and forwards against the studs – at worst chopping them off, and at best wearing the stud-holes oval and

159

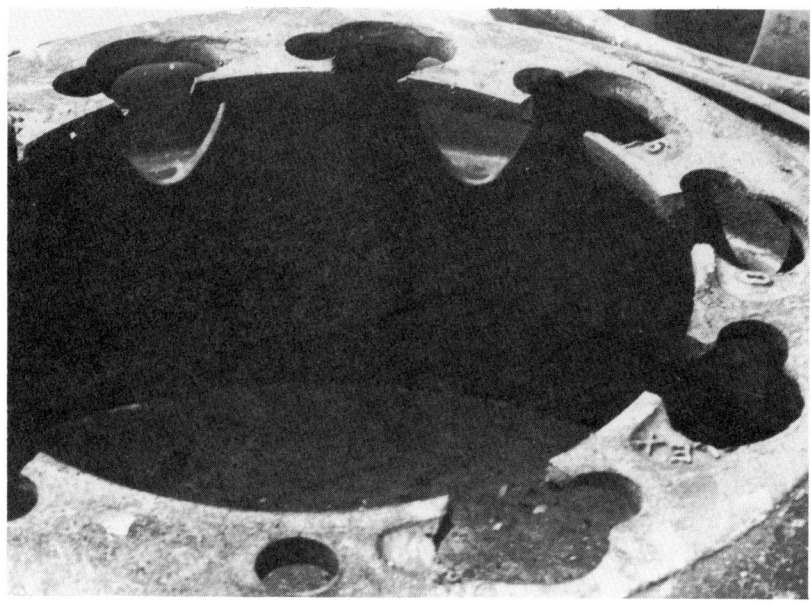

Figure 11.1 *Oval wear of holes of spigot wheel, followed by wheel breakaway.*

weakening the wheel. (In both fixing arrangements, looseness also brings shock end-loads on the studs.)

It is essential, therefore, to tighten nuts enough for the frictional forces in the threads to stop the nuts slackening. There is still a lack of precise evidence to indicate what tightening torque gives that required lock-up frictional force in the threads. Guidelines from the fastener manufacturers suggest, however, that it should produce between 70 and 80 per cent of yield-point strain on a 22mm stud. On that basis, most recommended torques are insufficient to keep nuts tight for long periods. This has lately been confirmed by research work by GKN.

The nut-torques often recommended by manufacturers develop about 8,000lbf (4,000kgf) clamping force with T-grade (55 ton) studs and 14,000lbf (6,400kgf) with V-grade (65 ton) studs. Neither is sufficient to resist maximum braking torque. Moreover, it is worth pointing out that with the more powerful vehicles lately introduced, the driving torque now being put through rear axles can be as great as the braking torques. This seems to explain why coaches and the tractors of top-weight articulated trucks are suffering particularly badly from wheel-security problems, and why recent cases have been observed with the holes of spigot wheels wearing oval in the driving rather than the braking direction.

Figure 11.2 *A fixed wheel cone-nut torn away after nut looseness developed.*

Despite the fact that overtightening encourages stud-stretch and failure (and compressed-air impact wrenches are prime offenders here) the general situation is that nut torques can be higher than is at present generally recommended.

The Institute of Road Transport Engineers, which has still not concluded a major study into the problem, has found that a big reason for limiting tightening torques on coned-nut assemblies was distortion of the slender section of wheel nave under each stud-hole. Permanent wheel distortion was feared if the tightening torque exceeded about 550Nm (400lb/ft) on dry threads. That is only the case with plain nuts and coned washers, however, which waste less torque in friction. With coned nuts the tightening torque could go up to 800Nm (about 600lb/ft). These figures were from tests with the wheel backing on cones; when the wheel was flat against the brake drum or hub the wheel distortion was very much less.

Taking all this into account, together with the strengths of the alloy steels used for studs and the need for extra clamping force on spigot wheels, I am inclined to suggest the following maximum tightening torques on dry (unlubricated) threads:

Coned nuts, 800Nm (550lb/ft)

Plain nuts with coned or spherical washers, 650Nm (450lb/ft)

Plain nuts with flat washers on spigot wheels, 900Nm (650lb/ft)

To lubricate threads seems a bad idea, because these tightening

Figure 11.3 *Wheel wear caused by oscillation of wheel nuts because of driving torque.*

torques could then overstress the studs and still not achieve the friction-force needed to keep the nuts tight.

To limit wheel distortion it seems most desirable never to fit wheels thinner than 12mm with coned fixings.

The tightening torque I recommend for spigot wheels is sufficient even to hold wheels with painted faces, whereas at present manufacturers often blame spigot-wheel shift on paint. No practical tightening torque will hold spigot wheels that have oil or grease on their faces, however. Lubricants on wheels should therefore be absolutely taboo.

There are plenty of other things that should be taboo as well, for just tightening enough to keep the nuts secure is no cover for careless fitting and wheels in bad condition. Good maintenance is still essential and basically that means cleanliness and inspection.

Wheel-faces must be clean, dry and free from any lubricant. Only a thin coat of paint should be allowed on the faces. If any grit is trapped the studs will be bent slightly, the wheel and maybe the brake drum will be distorted, corrosion will develop and clamping friction will be lost. It is likely that the nuts will not stay tight either.

When a wheel is on, keep an eye on stud stretch; this is easier to do if the stud starts life flush with the nut or has a groove in the end as deep as the nut. Look for bright marks, paint flaking or rust streaks around the nuts. Look, indeed, for any missing nuts or broken studs; as soon as one goes, more strain is put on the remaining ones, and the rest are inclined to suffer the same unhappy fate. Such visual checks are really a matter of just a slow walk round the vehicle; they should be part of a professional driver's rest-stop procedure, looking at tyres at the same time. Drivers should use their ears, too. Sometimes incredible distances are travelled with a loose wheel that must have been rumbling and knocking ominously for a long time. Whatever happens, these elementary inspections, together with checks of nut tightness with a torque wrench, should be part of the regular docking procedure.

If a vehicle comes home after having had a wheel changed on the road, it is advisable to take it off again, just to make sure the seating is clean, the wheel sound and the nuts tightened properly. Whenever a wheel is off the opportunity should be taken to check it for cracks or distortion. With spigot wheels beware of ovality in the holes or wear of the faces around them; and be on the lookout for sloppy fits over the spigot.

Ovality is a danger signal with holes in coned-fixed wheels, too. It produces offset loadings into the studs. There are several other things on which to keep an eye, though, with coned-hole fixings.

Be suspicious of one-side markings on the chamfers and on the studs' seating cones. They are a sign of the pitch circles of wheel and studs not being identical – and that has caused many a failure for, again, it bends the studs when the nuts are tightened. The same can happen if studs have not been rooted squarely.

Similar symptoms, but with haphazard bite-marking, indicate eccentric nuts – more common than one might think. Beware cheap nuts. Quality is very important with coned seatings. Check for sloppiness on the threads, run-out while the nut is rotated and the angle of the cone.

Studs themselves should be felt for looseness. Sometimes the depth of root leaves much to be desired and wear can develop; then, even though the nuts are tight, the wheel still wobbles on rocky studs.

Sometimes, with split cones over the studs, the cones grip the stud firmly enough to give the impression that the nut is tight, even though the head of the stud has broken. For this reason, solid cones are really preferable, and convert more torque into friction into the bargain. All loose cones are a bit of a liability, though, from a maintenance point of view. They are sometimes missed off or the wrong size is fitted – both disastrous.

Then there are the chamfers in the wheel-holes to watch. Evenly distributed wear, even if quite heavy, is more tolerable than is often stated, but uneven wear or some holes worn more than others – well, that is an

easy recipe for sour events. The seatings on the cones become uneven, that vital friction-hold is lost, the wheel is distorted and some studs are overloaded, while others do little. The plug gauge sold by Scammell and the vernier gauge sold by Shell are both useful pieces of checking equipment for chamfered wheel-holes.

Wheels that have been reconditioned need particularly careful checking. Even new wheels are not perfect sometimes.

Even after such diligent maintenance, one cannot guarantee the end of wheel-security worries. Designs can leave something to be desired, too. The fact that some makes and models suffer more than others from wheel-fixing defects points to differences in design.

Common deficiencies are studs threaded too far down and with the rooting too shallow. But problems can have more subtle causes. One is differential expansion of the pieces clamped together. The higher-tensile steel studs expand more than the iron and mild steel of the wheels, hubs and brake drums. When they all warm up together the nuts lose some tightness. When just the brake drum is hot, before it has soaked through, the studs are put in extra stress. They are likewise stressed if the whole clamped assembly goes very cold – the studs contract more than the rest. What can be critical is the relative thicknesses of the components in the assembly.

Use of left-hand threads on the left-hand side of axles has been found beneficial at the margin. Although the hand of thread cannot be shown to have any effect while nuts are tight, it does discourage looseness developing if torque is lost – when a spiralling pattern of stud deformation and wheel friction against nut begins to take place. No difference between left-hand and right-hand threads has been observed in the matter of proneness to stud breakage, however.

Clamp-torque comparison

Wheel nut	80% of yield stress N/mm^2	Corresponding clamp torques required (NM)		Clamp torques commonly recommended Nm
		DRY	LUBRICATED	
⅞in BSF T grade	510	850	680	380-550
⅞in BSF V grade	640	1080	865	400-620
22mm MF 9.8 grade	510	765	615	480-540
22mm MF 10.9 grade	640	970	780	550-690

Brake-drum flange 150°C hotter than rest of fixing assembly

Wheel nut	Drum expansion mm	Added clamp load kgf	Added stress N/mm^2	80% yield stress N/mm^2
10.9, MF spigot	0.305	3790	118	640
T, BSF cone	0.305	3490	117	500
V, BSF cone	0.305	3540	119	640

12. Commercial Vehicle Driving Techniques

Keith Parmee

The commercial vehicle's prime function has always been to reduce man's physical burden of goods. Steady evolution has seen the replacement of man as a beast of burden, by the harnessing of animals to provide the power source and, of course, in this century by the development and use of the internal combustion engine. This latter advance signalled a rapidly increasing development, during which time the commercial vehicle has evolved into a highly efficient load carrying machine. Man is no longer expected to provide the motive power, but still maintains control over the machine, and our existing social standards demand an increasingly more comfortable working environment. Modern trading conditions demand that journey lengths and/or traffic density increase. The physical and mental pressures on the driver must, therefore, be reduced to a minimum.

It is perhaps not readily appreciated that in driving, apart from steering the vehicle, gear changing is the most frequent energy using task which the driver has to perform. As cab comfort and simplicity of control are increased, and the use of electronics in vehicles is becoming more widespread, it is opportune for us to review past and present driving techniques and to explore some of the possible routes for future development, all aimed at reducing driver fatigue, allowing the driver to concentrate on traffic conditions, thereby increasing safety.

From an operator's point of view, any improvements made must give financial payback and must, if possible, reduce fuel consumption and down-time because of driveline durability problems.

The Past

With the commercial vehicle, the amount of power available from the motive source has increased dramatically. This has allowed the maximum gross weight of the vehicles to increase and allowed us to propel such higher weights at higher speeds. To accommodate these increases the torque capacities of gearboxes have increased as have the number of gear ratios and overall gear ratio spreads available.

We have seen the development from simple five speed units, through six, nine, ten, 13 and above. The increase in speeds was necessary in

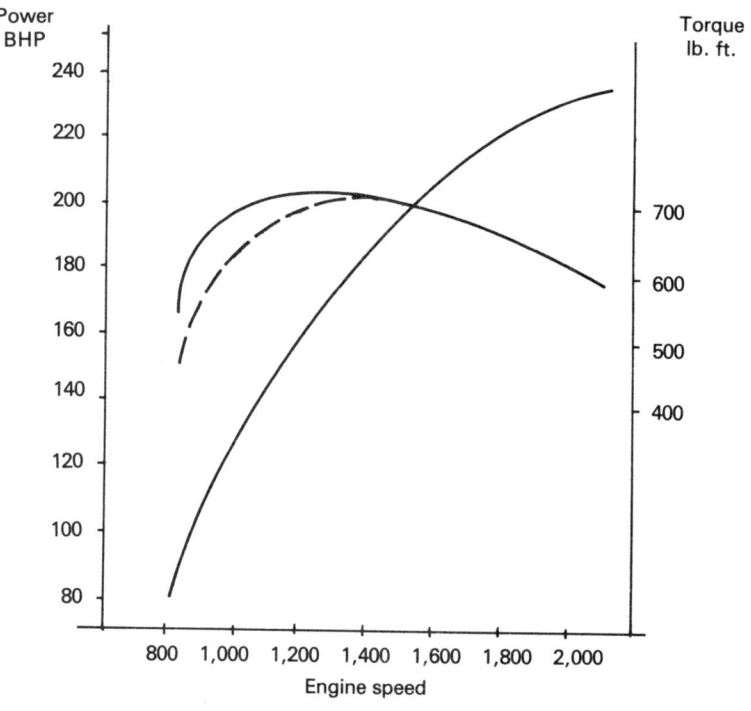

Figure 12.1 *Graph showing available torque contrasted with manufacturers' estimates*

order to allow greater overall ratio coverage, to maintain high climbing and restartability and because of higher vehicle weights and the use of high-power turbocharged engines.

The development of turbocharged engines has increased the maximum torque and power available but, depending on the matching, it can have a detrimental effect on the torque available at low engine speed. In addition, and in a restart mode, for example on a hill, the engine will be accelerated as the clutch is engaged, but this will not allow the turbocharger to have gained sufficient revs to have become effective. The torque actually available from the engine in this condition is substantially less than that shown in manufacturers' published data (see Figure 12.1). The result of this is to increase further the overall ratio coverage required from the gearbox.

Second, increasing pressure for fuel efficiency has necessitated the provision of smaller gear steps, enabling the engine to be operated within its most effective speed range.

In Europe, the first method of developing more spread and speed was to take the basic five speed unit and add to it a sixth gear, in the form of an overdrive, to the rear of the transmission (see Figure 12.2). This had the effect of increasing the ratio spread but not reducing the gear steps. The second method was to add a front-mounted splitter unit which would not only extend the ratio coverage but also split the main gearbox ratios to give smaller steps. This unit was usually air-operated and could be either synchronised or non-synchronised. There are, however, limitations with this concept; for example, in the overall ratio coverage that can be obtained and in the step size obtainable which can make the driving technique 'fussy'.

In the United States, the increase in ratio coverage was achieved by using an auxiliary gearbox which had to be shifted separately by the driver, usually by a separate shift lever. In the late 1950s, the Eaton Corporation perfected the design of making the auxiliary gearbox integral with the main gearbox and using pneumatic shifting for the auxiliary. This became known as the Roadranger, a design technique which has been followed for all units known as Roadranger gearboxes. The new design not only reduced driver workload but also enabled a greater overall spread than could be achieved by the front-mounted splitter alone. The reason for this is that in any main gearbox of the size used in commercial vehicles, there is a maximum gear reduction that can be achieved using two pairs of gears. In a front-mounted splitter, the step size has to be such as to evenly split the gears in the main box and this is usually in the order of 30 per cent. This then is the maximum extension that is possible to the ratio spread of the main gearbox.

A further disadvantage of the front splitter unit is that because the main gearbox ratio steps are usually arranged in a geometric progression for good driveability, the split step cannot equally split each gear. This leads to unusable ratios and, in some instances, confusing gate selections.

In the range unit, the gears in the front box are used twice, in both high and low range. Therefore, the overall reduction in low range is accomplished on four pairs of gears and thus a greater ratio coverage is possible. In this type of design the ratio steps in the front gearbox must be constant, which gives equal step sizes in both high and low range with obvious benefits in fuel economy.

In many parts of the world, an alternative to the front-mounted splitter as a means of extending the gearbox ratio coverage was the use of a two speed axle. This performed the same function, had a similar driving technique and had similar limitations to the splitter concept. Operation was by either pneumatic or electric shift motors controlled

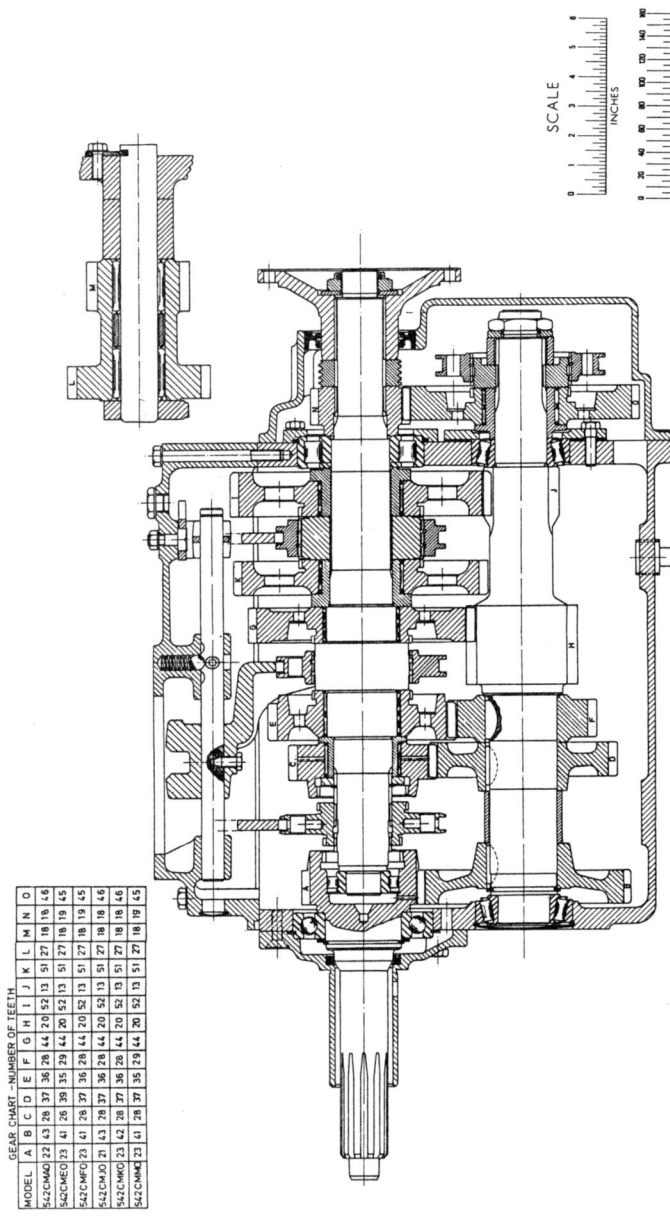

Figure 12.2 *One method of developing more spread and speed – the addition of a sixth gear or overdrive*

from the driver's cab and used throttle manipulation to achieve synchronisation.

Because of step size problems, mentioned previously, for front-mounted splitters, it was found that drivers would only use the splitting ability of the two speed axle on the top two gears – effectively turning a five speed transmission into a seven speed. They would, however, use the low range of the axle in first or second gear simply as a ratio extender, but would leave the axle in any one range until fourth or fifth before using it as a splitter.

Present

Present day gearboxes for heavy commercial vehicles are generally range-change units, although the front-mounted splitter does still exist. In fact, some designs incorporate range-change and front splitter, although it is questionable whether the very small steps so produced are really necessary. The 'engine step', that is, the ratio (max gov rpm/max torque rpm) to one, is now no longer the criterion for calculating transmission steps. Step size is now arrived at from a consideration of the 'engine economy step', that is the ratio (highest economy rpm/lowest economy rpm) to one, in the worst load conditions, while also taking into account the requirement for this step to fall within the usable torque range.

A study of modern engine trends shows this 'usable economy step' to fall mainly within the range 25 per cent to 40 per cent, with the larger intercooled engines tending towards the 25 per cent to 30 per cent range.

A general purpose, premium highway transmission for the 1980s, therefore, should preferably have ratio steps not greatly exceeding 30 per cent and, ideally, between 25 per cent and 30 per cent. Larger steps will reduce driveability, while smaller steps will increase the driver's workload. Coupling this with the ratio spread required to restart at maximum gross weight on 16 per cent grades and to achieve maximum legal level motorway speeds within the usable economy rpm band, indicates that 12 or 13 speeds may be the optimum – always remembering that the goal is to reduce the driver's task to a minimum.

Today's gearboxes, whether range or splitter units, have pneumatically operated shifts for the range or splitter, operated simply by a switch from the driver's cab. Some, however, have the pneumatics interlocked to the clutch pedal necessitating that the clutch be operated for every shift. This usually occurs with front-mounted splitter units whether applied on their own or to range-change units, and also to synchromesh units. The result is that the driver with this type of unit will have to work harder because the clutch must be depressed for every shift.

The other main design difference which exists today is between what has become known as 'synchromesh' and 'constant mesh'. Although these terms are convenient in differentiating the two systems, they happen to be misleading. All gearboxes today are constant mesh in that no design slides the gears into mesh; they all use constant meshing gears with sliding, clutching dog teeth. In addition, all gearboxes have to be synchronised before these clutches can be engaged. It is the manner of achieving this synchronisation which differentiates the designs. In what is known as synchromesh, friction elements are placed in the gearbox which speed up or slow down the gears, shafts and clutch spinner to achieve synchronisation. The energy necessary to do this is provided by the driver in the form of high shifting loads. In the so-called constant mesh gearbox this energy is provided by the engine, with virtually no effort required by the driver. Normally, with the constant mesh unit the engine would be used to accelerate the gearbox inertias for down-shifting and the double clutching technique would be used. However, the Eaton Corporation has developed the design of the clutching teeth so that considerable tolerance can be allowed on achieving synchronous speed with engagement still possible.

This has meant that professional drivers have perfected the driving technique of shifting the gearbox for all shifts without using the main clutch, a technique known as 'float shifting'.

There are, of course, advantages and disadvantages for both synchromesh and constant mesh gearboxes. The synchromesh gearbox does not require double-de-clutching, but does require use of the clutch for every shift and high shift lever loads (see Figure 12.6). Because of the high lever loads and the physical size of the synchronisers that are required, shift times are longer than the constant mesh, and by their very nature, such friction elements are wearing parts. The 'constant mesh' does not have these parts so must have longer life and have reduced weight. In addition, it has virtually no lever load but requires double-de-clutching unless the float shift technique is adopted.

It may at first be thought that, because the engine is being used to synchronise a 'constant mesh' gearbox, this is less fuel efficient than synchromesh.

Examination will, of course, reveal that this is not so because, first, at the time the engine is accelerated for down-shifts it is in a no-load condition and therefore uses very little fuel; also, although the engine is not accelerated to achieve synchronisation on a synchromesh, it must be accelerated after the shift, during clutch re-engagement, in order to match engine speed with vehicle speed in the new gear.

At present, apart from the main difference between synchromesh and constant mesh the other main difference in heavy truck trans-

Commercial Vehicle Driving Techniques

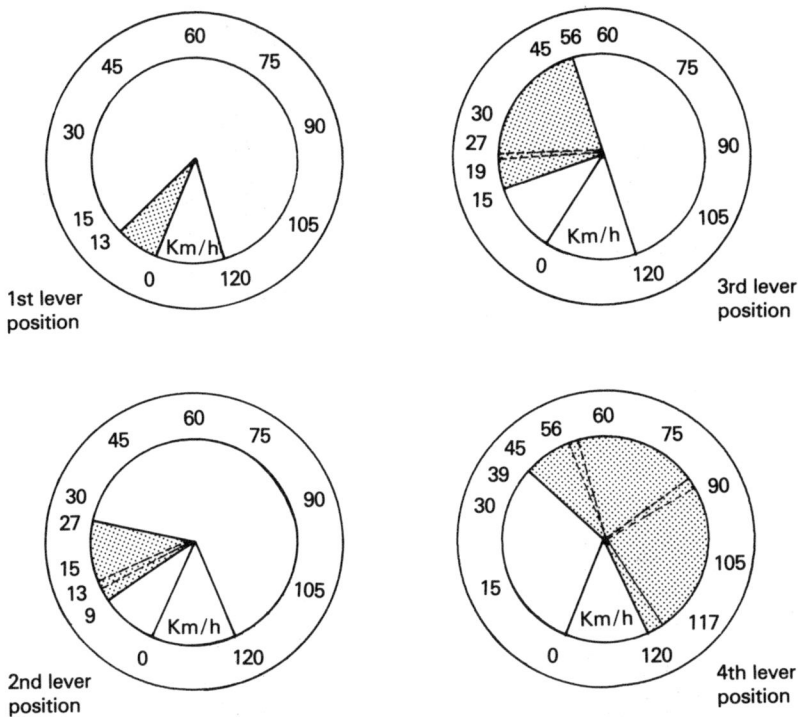

Figure 12.3 *Example of the speed usage of a 4×3 splitter*

missions is in the use of single layshaft or multi-layshaft systems. In practice, the benefits of multi-layshaft systems are really only apparent on units above 600lbft. torque capacity and anything over two layshafts causes technical and manufacturing problems which outweigh any benefits.

Eaton Corporation perfected the twin countershaft transmission many years ago and the benefits of reduced overall length and increased durability have stood the test of time in that over two million units have now been sold worldwide. The twin countershaft design allows the face-width of the gears and the gear tooth stress levels to be reduced. The floating mainshaft principle allows the gears to better align themselves to reduce further the dynamic stress levels and also to eliminate the need for bearings under the mainshaft gears as the loads on these gears are balanced. This reduces loaded parts, and thus potential problem areas, while at the same time making a much simpler transmission.

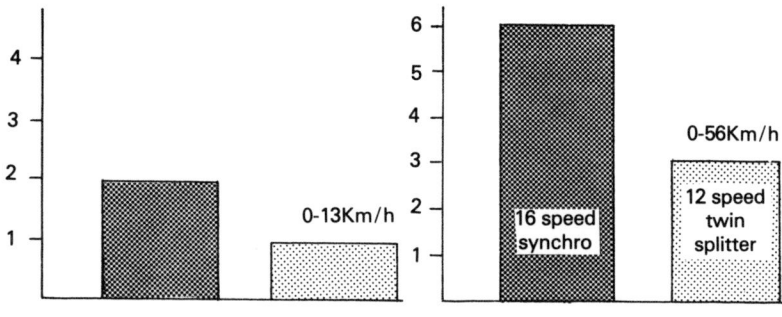

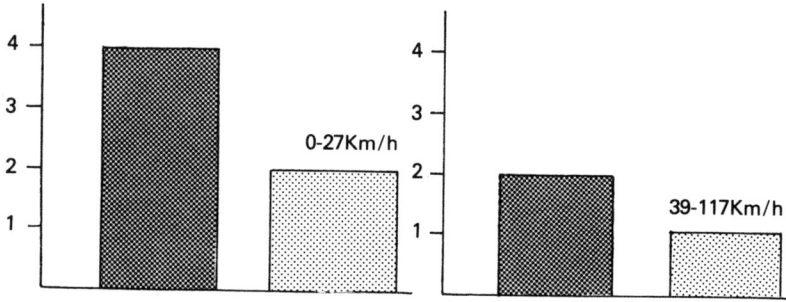

Figure 12.4 *Lever shifts*

Future

This part of the discussion is not intended to point out the course of development for any gearbox manufacturer, but is intended to indicate some of the possible courses of development which are available. Many of these are becoming more of a reality with the increasing use of electronics. First, it is evident that the driver's task must be made easier. Vehicle and engine trends will continue to demand multi-speed gearboxes, so methods must be found to reduce the physical effort of gear shifting. The synchromesh unit is already at a disadvantage in this and more use must be made of servo-assistance for shifting, even though this must inevitably add weight and cost. The need to reduce the gear shift lever load is witnessed by the fact that a number of synchromesh manufacturers have already announced servo-mechanisms. While these mechanisms will indeed reduce the driver effort, a possible

Commercial Vehicle Driving Techniques

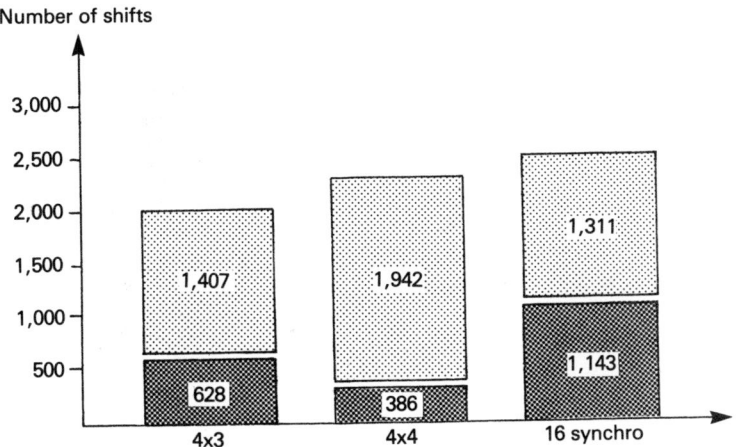

Figure 12.5 *Illustration showing the number of shifts on the road for 4×3, 4×4 and 16 synchro gearboxes*

side effect could be that overspeeding of engines will become more of a problem. With this development, the current synchromesh driving technique of using lever and clutch will not change; loads will simply be reduced. The constant mesh gearbox has already the advantage of low shift lever loads and can form the basis for some interesting future driving techniques.

Through the combination of a rear-mounted three or four speed splitter unit, a three or four speed main gearbox and the use of a sensing mechanism for shifting the splitter, a 12 or 16 speed gearbox can be produced, where the driver only moves the shift lever three times from rest to maximum speed and only once from 19 km/h to maximum speed. All other shifts are accomplished by operating a cab mounted switch and manipulation of engine speed. No clutch operation is necessary during the shifts. (Figures 12.3, 12.4, 12.5 and 12.6 compare this driving technique to the synchromesh one in terms of the number of lever shifts and clutch operations and in terms of the energy needed to shift, and demonstrate considerable reduction in these factors. (Lack of frictional wearing of parts and reduced clutch operation means longer life trans-

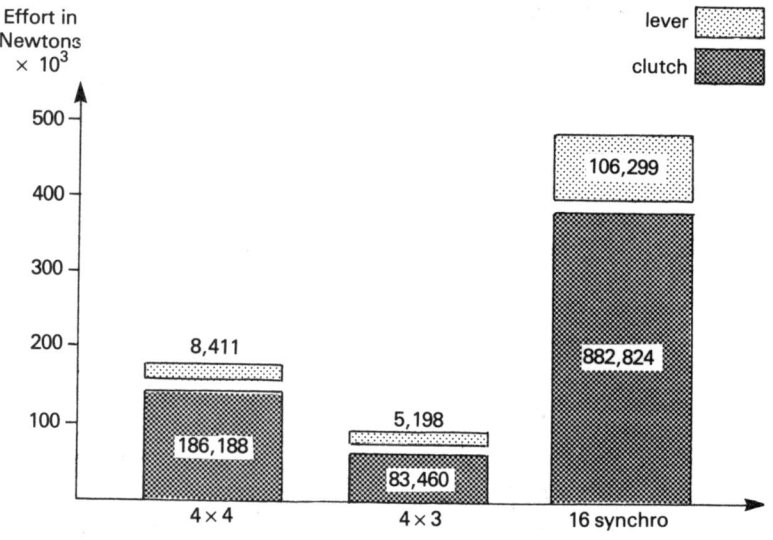

Figure 12.6 *Illustration showing physical driver effort for 4×3, 4×4 and 16 synchro gearboxes (NTC 300)*

missions. Because this gearbox also uses the proven long-life twin countershaft principle it is known as the Eaton Twin Splitter transmission.

By placing the splitter unit at the rear of the main gearbox we find that gearbox life is extended considerably. First, the front box unit spends most of its time in direct drive with no load in either gears or bearings. Second, the rear box gear, once designed to cope with first gear torque from the front box, spends most of its life with the front box in direct contact and thus at much lower torque levels. The unit's life is therefore increased substantially.

Through the use of electronics and the extended use of a sensing mechanism, a further transmission system can be produced whereby the in-cab shift lever and the associated change-speed mechanism between cab and gearbox are replaced by a smaller in-cab lever, effectively an electrical switch supplying signals to an electro-pneumatic actuation system mounted on top of the gearbox unit itself. The shift lever has only one plane of operation, fore and aft. Forward movement of the spring-loaded lever signals an upshift, while rearward movement signals a down-shift. Once signalled, the driver would simply break torque and raise or lower the engine speed by movement of the throttle. At synchronous speed, the gear would automatically engage so the clutch would only be necessary for initial vehicle starting. Skip shifting is accomplished by use of the button on top of the shift lever which can be used to signal as many up or down shifts as are possible within the engine speed range. Alternatively, the button can be dispensed with and by fast action on the in-cab lever the electronics will signal the down-changes so fast that only the last one is carried out. Once selected, the shift is carried out as before, by one movement of the shift lever and subsequent adjustment of engine speed (which may be by driver manipulation or performed automatically as an integral part of the control philosophy).

If throttle manipulation is carried out automatically, a pneumatic cylinder is connected to the fuel pump and signalled for 'blipping and dipping' from the electronic control system.

This driving technique gives no shift lever loads, no clutch use, except for starting, and numerous benefits to the vehicle builder in that the problems of running gearchange mechanisms around engines and turbochargers, having left-hand and right-hand changes are all eliminated, as too is the necessity for a hole in the cab floor through which gearchanges (and engine noise) usually pass. The end result should be a quieter and lighter vehicle. This may be the most that can be achieved without full automation.

The ultimate transmission system which gives reduced driver effort, both physical and mental, must be a fully automatic gearbox. Although fully automatic gearboxes have been available for many years, their use has been limited because the types available so far have proved expensive, have a high fuel consumption and are unreliable. These units have always used torque converters and usually have planetary gearing.

With the technology now available in electronics, it is possible to control a standard mechanical layshaft gearbox and clutch to obtain a fully automated unit which is fuel efficient and has the reliability and serviceability of commonly encountered mechanical transmissions.

Such transmissions are currently under development and, in the case of Eaton, can use the standard Roadranger gearbox and a dry plate clutch. The system has no clutch pedal, only accelerator and brake. All gearshifting and clutch actuation is automatically controlled, including the starting mode. Shift points are determined by road speed and throttle position. Such systems must be equipped with a manual override which allows the driver to hold any gear and upshift or downshift independent of normal predetermined speeds. Engine protection to this condition is essential.

The advantages of such a fully automatic unit are not only the reduced driver effort but also the fuel savings that can be obtained by programming the control logic to operate the engine always at its most efficient point. Because there is less driveline shock with a fully automatic gearbox there will be less driveline abuse and therefore increased durability of gearbox, drive-shaft and axle components. Any transmission system which can eliminate the need for a mechanical speed change control will eliminate the need for such a mechanism to protrude through the cab floor and thus will enable cab designers to reduce in-cab noise. An automatic gearbox based on this principle must be a more cost-effective solution than any of the previously tried concepts in heavy trucks.

The overall effect of such automatic gearboxes will be to reduce driver fatigue, reduce fuel consumption, reduce vehicle down time and increase vehicle safety.

The evolution of today's heavy commercial vehicle has meant that while the driver is now enclosed in a more comfortable and ergonomically-designed cab with power steering and servo-assisted clutches and brakes, he has to do more gear changing than ever before. This has been brought about not only because of more speeds, but also by denser traffic conditions and the need to operate today's engines in narrower speed bands to maintain fuel efficiency. The time has now come for us to pay more attention to the gear changing function and to devise methods of cost-effectively reducing driver fatigue. Nor should we accept today's position as the ultimate development and we must be ready to adopt shifting techniques which may be a little different from what we are used to, but which can offer substantial benefits.

13. Down Licensing and Down Plating

John Beveridge

Make no mistake, the new Transport Secretary, Nicholas Ridley, has inherited the most contentious 'hot potato' concerning road transport since his predecessor-but-one, David Howell, took the decision to permit a weight increase to 38 tonnes early in 1983.

This concerns the industry's continued quest for the introduction of Down Licensing, a scheme to allow operators to tax their vehicles in accordance with the weights at which they actually run. This would do away with the current vogue for Down Plating, a method which achieves the same end, but which requires costly and time-consuming physical alterations to each vehicle for acceptance.

This inheritance is made all the more unenviable by the Government's apparent unwillingness to legislate over the issue, a reluctance which has passed it into the hands of the third transport administration since its conception, and a delay which has resulted in even greater confusion.

The relative merits of Down Licensing as against Down Plating depend largely upon your position in this industry. Operators are for it; the Department of Transport appears to be against it; while manufacturers strongly advocate it – although their lives would be made much easier by a clear mandate either way. Whatever one's viewpoint, let us have another look at exactly what this issue involves – for it is only through a clear and thorough understanding of the relative pros and cons that any real sense of the issue can be made.

It has always been a prime objective of the Road Transport Industry to minimise its taxation burden whenever possible, taxation being a cost which is unavoidable. Likewise, it has always been the objective of the Department of Transport to maintain a fair and equitable taxation system for all concerned, while meeting the budgeted revenue targets required from it by the Treasury. Such a taxation system should also reflect accurately the true road track costs of each type of commercial vehicle, the main method upon which taxation scales are based.

Until the Autumn of 1982, these conflicting objectives lived in relative harmony alongside each other, under the old system of taxation according to unladen weight. Granted, there were certain pecularities and anomalies under this system, not the least of which was the difficulty in taxing certain vehicles, such as those in rental fleets, honestly. However, the system was as successful as could be expected with two such diametrically-opposed parameters – and it was easy to use, easy to administer and easy to police.

Unfortunately, it was also easy to abuse. A Government checkweigh in 1980 revealed that some 36 per cent of vehicles could be expected to have higher unladen weights than those declared for taxation purposes. Neither did it reflect accurately the true respective road costs of each type of vehicles; with some lighter vehicles allegedly paying proportionately more than their fair share of tax against their heavy articulated counterparts. It was against this background that the new Gross Vehicle Weight Taxation System was introduced in October 1982, which linked the level of tax paid to plated gross vehicle weight.

However, the advent of two other major changes in road transport legislation – the introduction of a 38 tonne weight level for five axle artics, and the introduction of Type Approval – are proving a major complication to the success of this GVW taxation scheme.

A higher weight level for artics has meant that those wishing to purchase and operate higher-powered vehicles, inevitably with higher potential plated weights, have to pay higher tax levels, regardless of whether they wish to take advantage of greater potential payloads or not. This, in particular, penalises those operators who carry light but bulky loads, who can never use anything like the maximum weight allowable, but who need higher-powered vehicles to ensure operational performance and flexibility. As such, with the historical tendency in the UK for operators to buy vehicles more powerful than that necessary for their basic requirements, GVW taxation is proving a major penalty – a penalty which was made all the more painful by the vastly-increased taxation rates introduced in the Budget in April, together with a change in the relative 'steps' in weight taxation to reflect the higher road track costs of heavier vehicles with fewer axles.

The introduction of Type Approval has had the effect of limiting possible reaction to this problem by manufacturers and operators alike. Since new vehicles must now be 'approved' to well-defined specification standards, any attempted changes to vehicle specification to counteract the relative penalties of GVW taxation would need to be costly and time-consuming, with no guarantee of acceptance by the governing body, the Department of Transport.

As such, two options become apparent: Down Licensing and Down Plating. Down Licensing is a scheme whereby an operator would declare the maximum weight at which he wishes to operate, and pay tax accordingly. In effect, he is promising not to operate above that weight with that particular vehicle, and to do so would be committing a Vehicle

Excise Duty offence. Down Plating is an exercise by which a physical change is effected to a vehicle to lower its plated GVW, allowing the operator to pay tax at that lower rate – it being physically impossible to use that vehicle at any higher weight without contravening Construction and Use Regulations, as well as committing a Vehicle Excise Duty offence.

All of the concerned parties in the industry recognise that Down Licensing is the more preferable of these two options. Indeed, a provision was included in the 1982 Finance Act to enable the Secretary of State to introduce Down Licensing, if this was felt necessary. It would allow operators to pay tax directly in line with the use that their vehicles make of the road, and, indeed, directly in line with the actual road track costs they inflict, the prime criterion of taxation assessment. The Government, however, continues to resist this scheme for a number of reasons. To begin with, it is felt this would be more difficult to police and to enforce. Policing under the current scheme involves a visual check of the vehicle's tax disc, plate and number of axles. With Down Licensing, a checkweigh would also become necessary. With the Government's stated intention for wider use of dynamic weighbridges, this argument does seem rather anomalous – particularly if incorporated with an individual, easily-distinguished tax disc scheme to identify stated GVW. If an operator was stopped and found to be operating above his stated taxation-weight, a penalty would be levied upon him commensurate with the offence he had committed. Supporters of Down Licensing would be quite happy for these penalties to be set at a sufficiently high level to deter abuse. Down Licensing would also result in a fairer distribution of the taxation burden on the transport industry and would not lead to a lowering of the overall level of revenue, as tax rates would be adjusted to take account of any potential loss through operators taking advantage of Down Licensing. It is advocated by the SMMT, FTA and RHA; it is supported by operators and manufacturers alike; and even Department of Transport Inspectors tend privately to view it as the best solution.

Down Plating tends to be viewed with considerable caution by all, however – even though all are being forced to take advantage of its provisions to achieve stated ends. However this issue is viewed, to achieve a lower plated weight on any vehicle requires a detraction from its general level of specification. A lower level of specification – whether achieved through changes to springs, brakes, propshafts, chassis or other driveline configurations – must, in essence, detract from vehicle safety and efficiency.

We are all aware that today's commercial vehicles are built generally to the highest standards. This has been the result of many years of continuous legislation to ensure that this is so, combined with demands from industry for an efficient transport system. Environmentally-aware manufacturers tend to offer, and efficiency-minded operators tend to buy, vehicles with considerable power and safety margins to ensure that maximum service is given to the consumer, through minimum disruption to his daily life. By forcing manufacturers and operators alike to Down Plate vehicles, surely the Government is acting completely in opposition to these stated objectives, when another essentially bureaucratic method of achieving the same end exists, if they bother to make use it?

The conclusion reached by any aware party must be that Down Licensing is infinitely preferable to Down Plating in whatever format. We, at DAF Trucks, believe this. We have advocated its introduction very strongly, and we shall continue to do so through all available channels.

But certain facts must be faced. At present, reasonable taxation levels can only be achieved through the Down Plating of vehicles to a level commensurate with the operating weight involved. As such, how have we – as a major tractor manufacturer – achieved this end?

Thorough market research concluded that three weight categories were to be of prime importance in the post-1 May market. To begin with, the newly-introduced 38 tonnes – and this was more than adequately catered for by existing model-offerings in power ranges from the medium level (244bhp) up to a reasonable maximum level (330bhp), particularly for operational cost levels. The two other weight categories identified were: 32.5 tonnes (the old 32 tons Imperial weight) and 28 tonnes.

With the traditional importance placed on the 32.5 tonne level in the UK, it was certain that many operators would continue to have their base loads linked to this weight level, until such time as their customers required them to carry greater payloads. The imposition of new safety regulations on vehicles above this plated weight-level, regardless of actual operating weight, only served to further emphasise this importance.

The 28 tonne sector has always been a category in the UK market, along with 24 tonnes, although of minor importance – as 28 tonne operators tended to purchase vehicles of 32.5 tonnes capacity to ensure operational benefits, there being no penalty for doing so. With considerable taxation savings to be made, even over and above 32.5 tonne levels, this could no longer be the case. Those operators who can physically never gross more than 28 tonnes (and this includes many national companies with large fleets) would continue to require high-powered vehicles, although now with 28 tonnes plated weight.

Twenty-eight and 32.5 tonne weight-levels could only be attained through Down Plating. This could be achieved through numerous methods: by modifications to brakes or suspensions, or by changes to drivelines. However, having recognised this, DAF Trucks were determined that any such Down Plating would only be carried out if it could be done without detracting from overall safety margins and without lowering any of the general strength and design parameters of DAF Trucks. We market quality commercial vehicles and, as such, could not agree to any change in specification which might be construed as diminishing our reputation for this.

Engineering research suggested that the optimum method of achieving Down Plated weight-levels would be to modify the fifth-wheel mounting flitch plates to levels compatible with both 28 and 32.52 tonnes – a method which would allow the specification of the vehicle itself to remain unaltered. These specially designed flitches were verified by computer analysis to ensure their general operational acceptability and were designed especially to suit current DAF 4 × 2 tractor models of whatever power – thus giving the operator totally flexible choice as to the vehicle he wishes to buy for his operation. This form of Down Plating also offered the DAF operator the opportunity to Up-Plate/Down Plate his vehicle at will, under the recognised VTG10 procedure, at any later date.

These Down Plated models were consequently submitted for Type Approval and, following consideration of computer analyses, drawings and vehicle identification documents, were accepted under Type Approval for operation in the UK.

Having followed the Down Plating road, we have taken considerable steps to ensure that this system is not abused. We still retain the fear, however, that – as time goes by, as is always the case – someone will find a loophole, or will abuse the system, perhaps abusing overall vehicle specification; a situation we have gone to considerable lengths to avoid. This can only be the result of the Government's decision not to allow Down Licensing, forcing Down Plating upon manufactuers.

Type Approval is a well-defined engineering system to ensure that vehicles operating on today's roads meet certain EEC directives, thereby making all commercial vehicles safer, cleaner and more efficient – all of which makes for a more environmentally-acceptable transport system. We believe in Type Approval; we believe in its concepts and we support it wholeheartedly. On the premise that it is applied primarily to standard vehicles, the system can only be to everyone's benefit. We do not, however, believe in de-specifying vehicles to achieve lower taxation rates. To do this would appear to be misuse of a system – especially when the same ends could be achieved through an administrative channel, a channel which has been provided for, and a channel which all participants within the industry itself would seem strongly to support.

In conclusion, it is surely far easier, far safer and, ethically, far more sensible to tax vehicles on their operational weights and control them in this way. This would ensure a system of taxation fair to all and which would allow the Treasury to meet its budgeted revenue targets and which could not be open to long-term abuse in terms of vehicle specification. We have gone a long way down the road towards an efficient transport system for the 1980s – let us have Down Licensing, and let us ensure that we stay on that road; do not force a U-turn upon us for purely bureaucratic reasons.

Appendix: Vehicle Weights and Dimensions

TWO AXLE RIGIDS

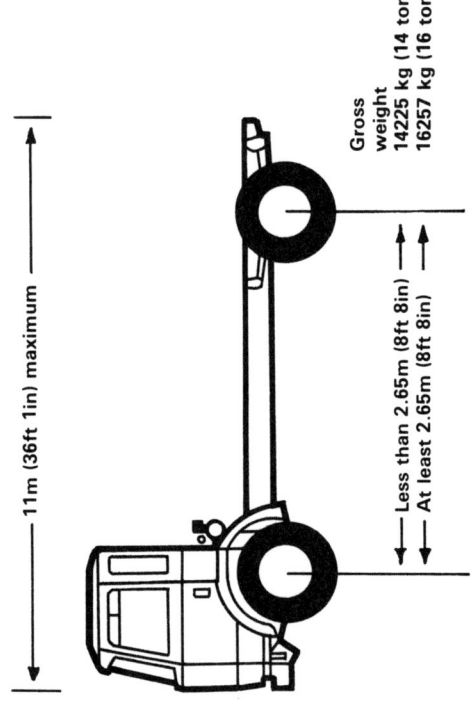

Appendix: Vehicle Weights and Dimensions

THREE AXLE RIGIDS

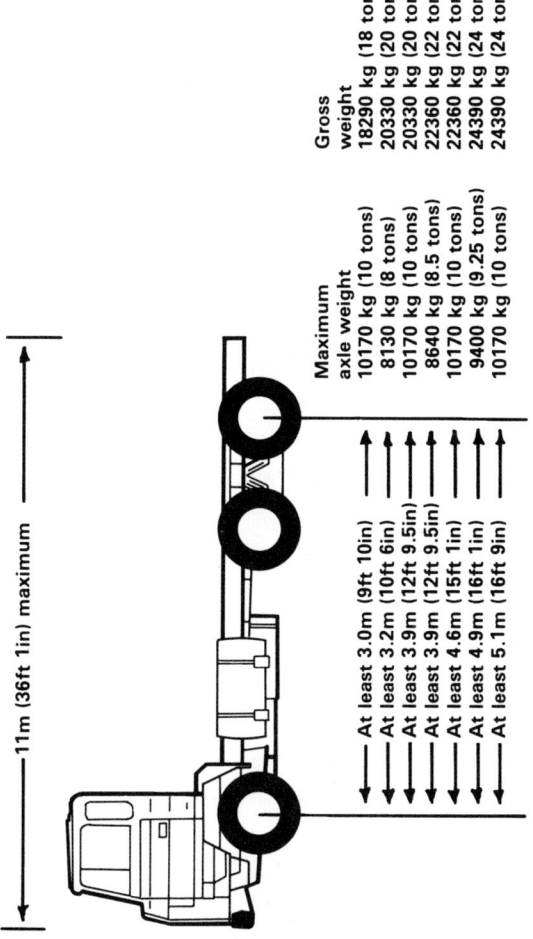

11m (36ft 1in) maximum

Maximum axle weight	Gross weight
10170 kg (10 tons)	18290 kg (18 tons)
8130 kg (8 tons)	20330 kg (20 tons)
10170 kg (10 tons)	20330 kg (20 tons)
8640 kg (8.5 tons)	22360 kg (22 tons)
10170 kg (10 tons)	22360 kg (22 tons)
9400 kg (9.25 tons)	24390 kg (24 tons)
10170 kg (10 tons)	24390 kg (24 tons)

At least 3.0m (9ft 10in)
At least 3.2m (10ft 6in)
At least 3.9m (12ft 9.5in)
At least 3.9m (12ft 9.5in)
At least 4.6m (15ft 1in)
At least 4.9m (16ft 1in)
At least 5.1m (16ft 9in)

187

FOUR AXLE RIGIDS

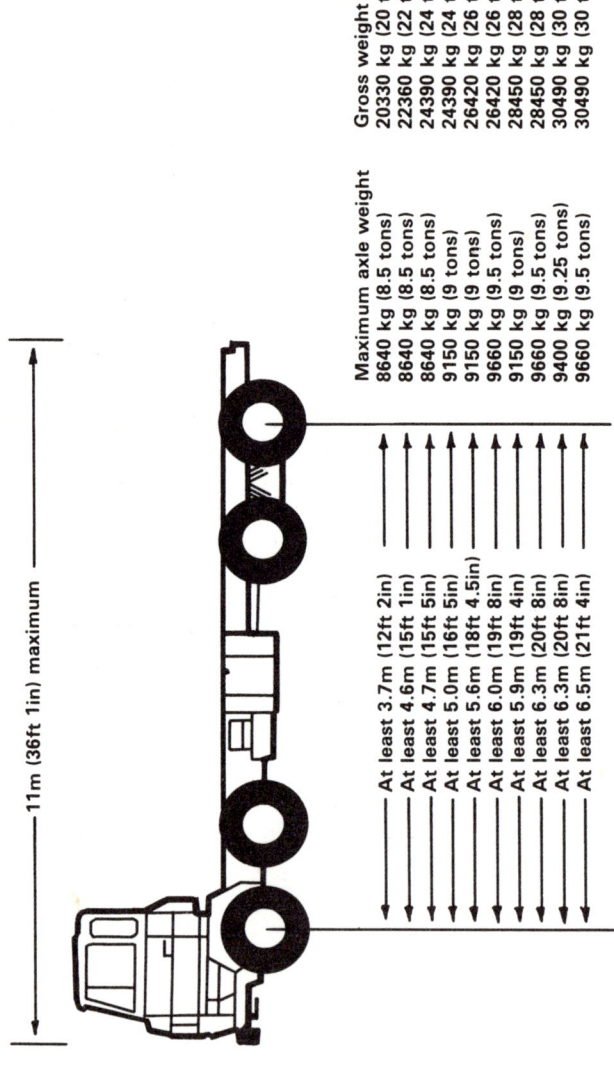

11m (36ft 1in) maximum

Maximum axle weight	Gross weight
8640 kg (8.5 tons)	20330 kg (20 tons)
8640 kg (8.5 tons)	22360 kg (22 tons)
8640 kg (8.5 tons)	24390 kg (24 tons)
9150 kg (9 tons)	24390 kg (24 tons)
9150 kg (9 tons)	26420 kg (26 tons)
9660 kg (9.5 tons)	26420 kg (26 tons)
9150 kg (9 tons)	28450 kg (28 tons)
9660 kg (9.5 tons)	28450 kg (28 tons)
9400 kg (9.25 tons)	30490 kg (30 tons)
9660 kg (9.5 tons)	30490 kg (30 tons)

- At least 3.7m (12ft 2in)
- At least 4.6m (15ft 1in)
- At least 4.7m (15ft 5in)
- At least 5.0m (16ft 5in)
- At least 5.6m (18ft 4.5in)
- At least 6.0m (19ft 8in)
- At least 5.9m (19ft 4in)
- At least 6.3m (20ft 8in)
- At least 6.3m (20ft 8in)
- At least 6.5m (21ft 4in)

Appendix: Vehicle Weights and Dimensions

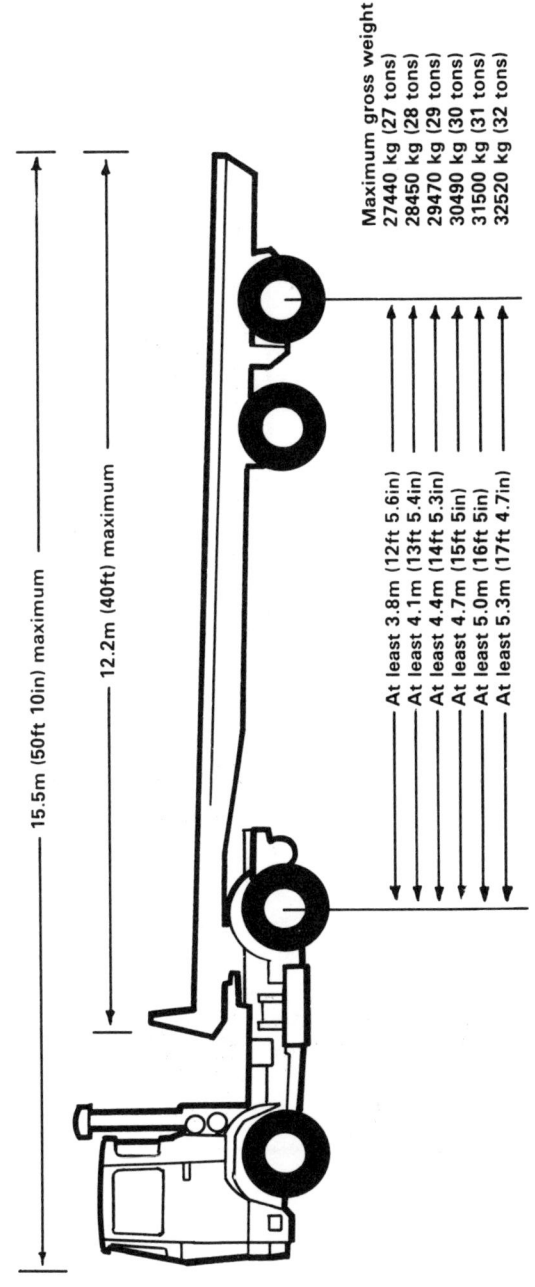

TWO AXLE TRACTIVE UNIT WITH TWO AXLE TRAILER

TWO AXLE TRACTIVE UNIT WITH SINGLE AXLE TRAILER

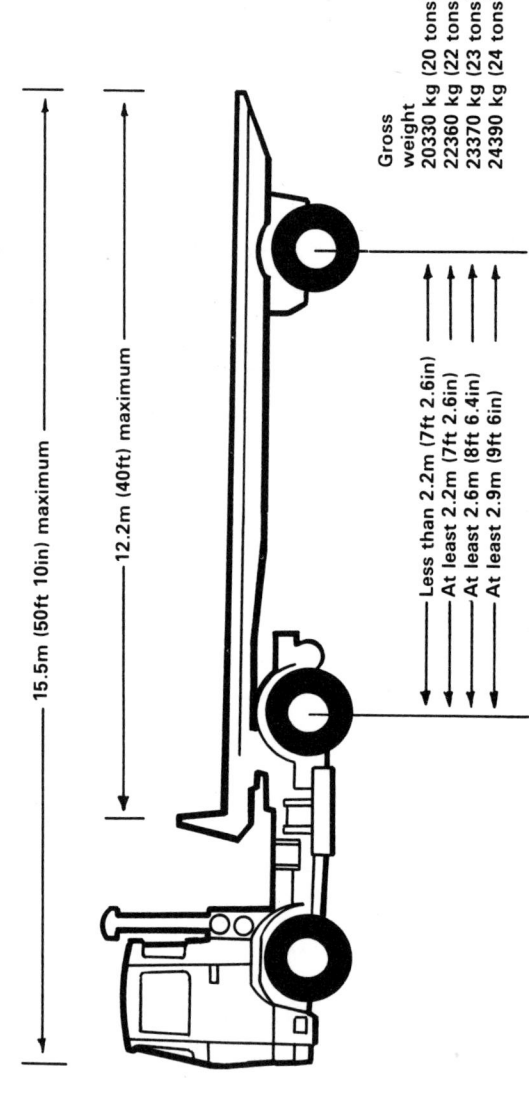

Appendix: Vehicle Weights and Dimensions

THREE AXLE TRACTIVE UNIT WITH TANDEM AXLE TRAILER

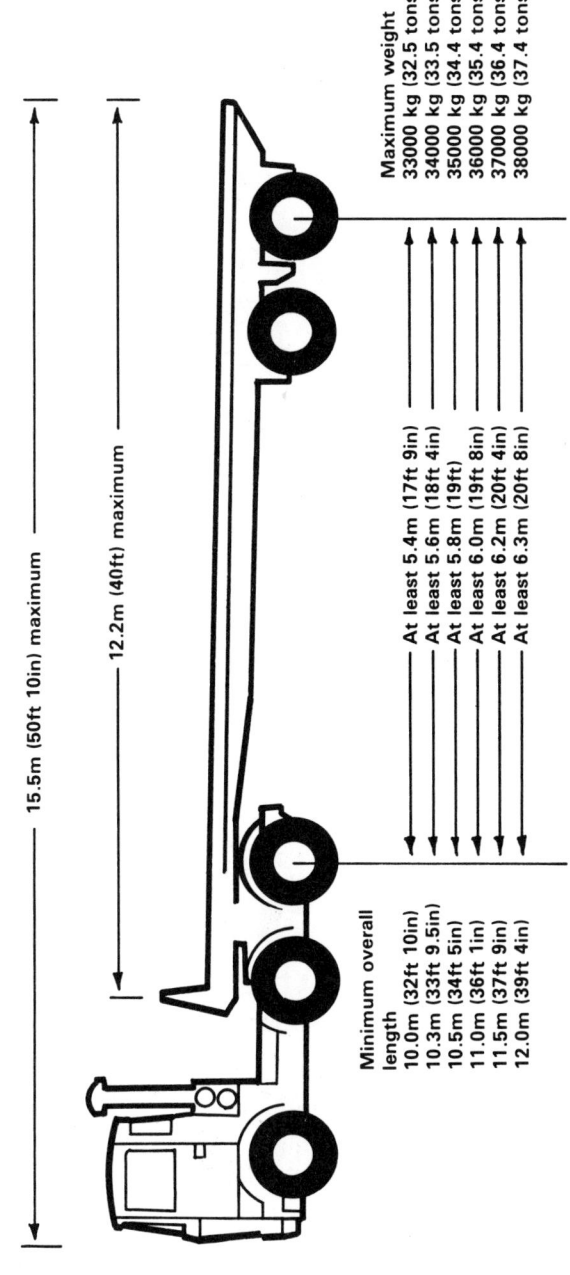

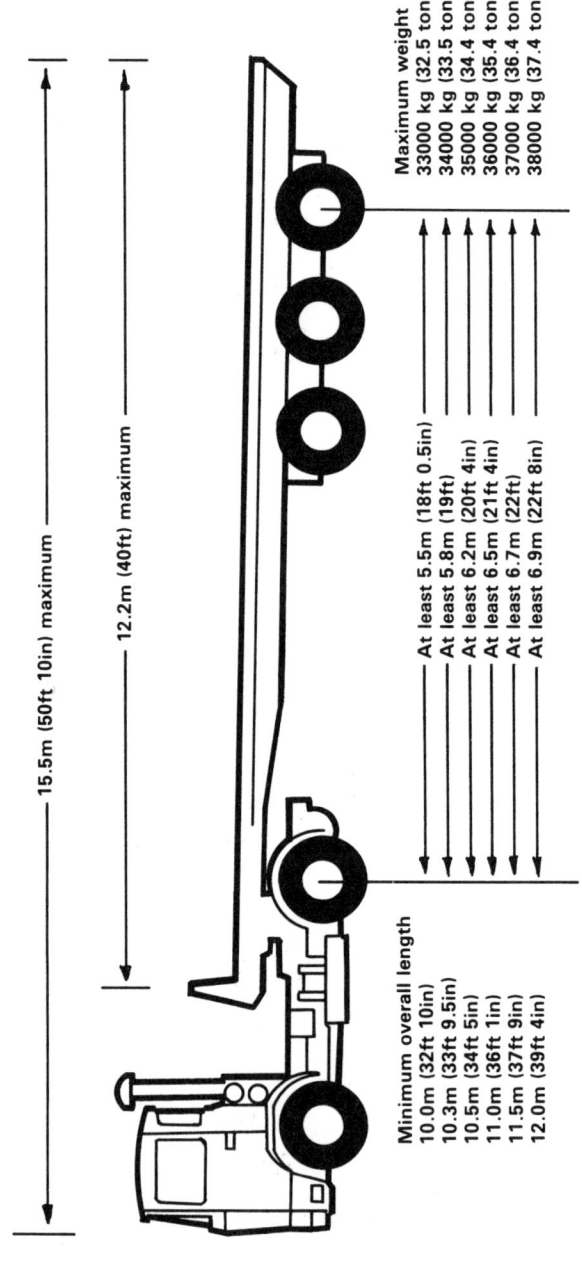

Appendix: Vehicle Weights and Dimensions

DRAWBAR COMBINATIONS

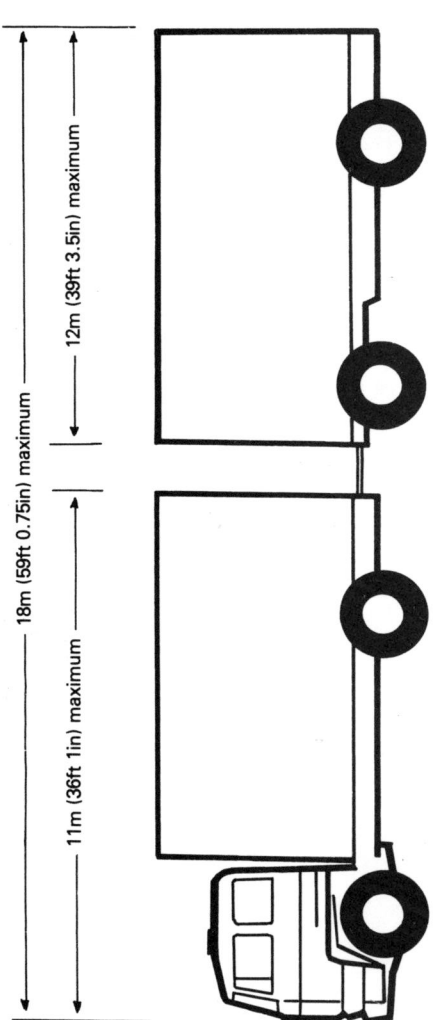

TWO AXLE BOGIES

At least
1.02m (3ft 4in)
1.05m (3ft 5in)
1.20m (3ft 11in)
1.35m (4ft 5in)
1.50m (4ft 11in)
1.80m (5ft 11in)
1.85m (6ft 1in)

(a)
Maximum weight when plated weight of neither axle exceeds one half of the specified weight
16260 kg (16 tons)
17280 kg (17 tons)
18300 kg (18 tons)
18800 kg (18.5 tons)
19320 kg (19 tons)
20000 kg (19.7 tons)
20340 kg (20 tons)

(b)
Maximum weight in cases not within column (a) when plated weight of neither axle exceeds 10170 kg (10 tons)
12200 kg (12 tons)
15260 kg (15 tons)
16270 kg (16 tons)
17280 kg (17 tons)
18300 kg (18 tons)
19000 kg (18.7 tons)
19320 kg (19 tons)

(c)
Maximum weight in cases not within (a) or (b)
10500 kg (10.3 tons)
10500 kg (10.3 tons)
15260 kg (15 tons)
16500 kg (16.2 tons)
18000 kg (17.7 tons)
19000 kg (18.7 tons)
19320 kg (20 tons)

THREE AXLE BOGIES

At least 1.4m (4ft 7in)
At least 1.6m (5ft 3in)
At least 1.8m (5ft 11in)
At least 2.0m (6ft 6.7in)
At least 2.2m (7ft 2.6in)
At least 2.4m (7ft 10.5in)
At least 2.7m (8ft 10.3in)

Maximum individual axle weight
6000 kg (5.9 tons)
6200 kg (6.1 tons)
6400 kg (6.3 tons)
6600 kg (6.5 tons)
6900 kg (6.8 tons)
7100 kg (7.0 tons)
7500 kg (7.38 tons)

Part 2: Directory of Manufacturers, Distributors and Trade Associations

Directory of Manufacturers, Distributors and Trade Associations

Axle Weighing Systems

Colindale Equipment,
Unit 7,
The Hyde Industrial Estate,
Colindale,
London NW9
tel: 01-200 7373

Loadax,
Probe Engineering Ltd,
Phoenix Way,
Cirencester,
Gloucestershire.
tel: 0285-67801

Precision Loads Ltd,
Craven House,
Carlton New Road,
Skipton,
North Yorkshire.
tel: 0756-69475

RAVAS UK Ltd,
43 Essex Street,
London WC2
tel: 01-353 0738

Smiths Industries
Automotive Instrument Systems
Ltd,
50 Oxgate Lane,
Cricklewood
London NW2 7JB
tel: 01-452 3333

Weighmax,
Hinckleys,
Tangier,
Taunton,
Devon.
tel: 0823-76041

Weightline Ltd,
33 Bouverie Square,
Folkestone,
Kent LT20 1BA
tel: 0303-42311

Weighwrite Ltd,
49 West Street,
Farnham,
Surrey GU9 7DX
tel: 0252-711011

Braking Equipment

Anti-Skid Controls Ltd (Maxaret),
367 Bedworth Road,
Longford,
Coventry CV6 6BN
tel: 0203-364777

Ashanco Autoparts Ltd,
35 Boldmere Road,
Sutton Coldfield,
West Midlands B73 5UY
tel: 021-355 3501

Belaco Ltd,
Chapel-en-le-Frith,
Stockport,
Cheshire SK12 6QE
tel: 0298-812545

Bendix Ltd,
Douglas Road,
Kingswood,
Bristol BS5 2NL
tel: 0272-671881

197

Berg Manufacturing (UK) Ltd,
Fountain Mill,
Carluke Street,
Blackburn,
Lancashire BB1 3JR
tel: 0254-59771

Brake Linings Ltd,
Bridge Street,
Buxton,
Derbyshire SK17 6BU
tel: 0298-77634

Clayton Dewandre Co Ltd,
90 Newbold Road,
Rugby,
Warwickshire CV21 2NL
tel: 0788-74561

Don International Ltd,
Hendham Vale,
Manchester M19 1SX
tel: 061-205 2371

Eaton Ltd,
PO Box 11,
Worsley Road North,
Worsley,
Manchester M28 5GJ
tel: 0204-72111

Ferodo Ltd,
Chapel-en-le-Frith,
Stockport,
Cheshire SK12 6JP
tel: 0298-812520

Hope Technical Developments Ltd,
High Street,
Ascot,
Berkshire.
tel: 0990-24855

Jake Equipment (UK) Ltd,
Claylands Close,
Dukeries Industrial Estate,
Worksop,
Nottinghamshire S81 7DW
tel: 0909-474848

Lucas Girling Ltd,
Kings Road,
Tyseley,
Birmingham B11 2AH
tel: 021-706 3371

Mintex Ltd,
Cleckheaton,
West Yorkshire BD19 3UJ
tel: 0274 875711

Perrot Brakes (UK) Ltd,
15 Upper King Street,
Leicester,
Leicestershire LE1 6XF
tel: 0533-544917

Retro Exhaust Brakes,
Kingsmead House,
Lyon Road,
Hersham Trading Estate,
Walton-on-Thames,
Surrey KT12 3PU
tel: 09322-25379

SAB Automotive Co Ltd,
Hilton Road,
Aycliffe Industrial Estate,
Newton Aycliffe,
County Durham DL5 6SX
tel: 0325-310110

Telma Retarder Ltd,
2/6 William Street,
Windsor,
Berkshire.
tel: 07535-51347

Trist Draper Ltd,
816 Bath Road,
Bristol BS4 5LH
tel: 0270-777093

Chassis Lubrication

Interlube Systems Ltd,
Estover Road,
Plymouth,
Devon PL6 7PS
tel: 0752-775781

Lubrication Equipment Ltd,
Lees Road,
Kirkby Industrial Estate,
Liverpool L33 7UD
tel: 0515 481183

Romatic Ltd,
Unit 3,
Derwent Street Trading Estate,
Salford,
Manchester.
tel: 061-834 0727

Telehoist Ltd,
Manor Road,
Cheltenham,
Gloucestershire GL51 9SH
tel: 0242-21355

Cranes

Associated Industrial and
Mechanical Engineers Ltd,
Goodley Works,
Colne Road,
Oakworth,
Keighley,
West Yorkshire
tel: 0535-43821

Atlas Hydraulic Loaders Ltd,
Wharfedale Road,
Euroway Estate,
Bradford,
West Yorkshire BD4 6SE
tel: 0274-686827

BCT Equipment Ltd,
Adam House,
58/66 Birmingham Road,
Kidderminster,
Worcestershire DY10 2SH
tel: 0562-740555

HIAB,
George Cohen Machinery Ltd,
23/25 Sunbeam Road,
London NW10 6JP
tel: 01-965 6588

Lyka Cranes Ltd,
382 Blackpool Road,
Preston,
Lancashire PR2 2DS
tel: 0772-727927

Palfinger,
PO Box 35,
Larbert,
Stirlingshire FK5 4NQ
tel: 0324-556222

Viper Cranes Ltd,
322 Short Heath Road,
Erdington,
Birmingham B23 7AN
tel: 021-384 4664

Demounts

Abel Demountable Systems Ltd,
The Priory,
High Street,
Redbourn,
Hertfordshire AL3 7LZ
tel: 058285-2685

Cartwright Freight Systems Ltd,
Atlantic Street,
Altrincham,
Cheshire WA14 5DH
tel: 061-928 2988

Kalmar Kockum Ltd,
Boyne Valley Industrial Estate,
Maidenhead,
Berkshire.
tel: 0628-39944

Ray Smith Demountables Ltd,
Botolph Bridge,
Oundle Road,
Peterborough,
Huntingdonshire.
tel: 0733-63936

Drive Lines

Borg & Beck,
Automotive Products plc,
Banbury,
Oxfordshire.
tel: 0295-4421

Brockhouse Ltd,
Hill Top,
West Bromwich,
West Midlands B70 0SN
tel: 021-556 1241

Cummins Engine Company Ltd,
46/50 Coombe Road,
New Malden,
Surrey KT3 4QL
tel: 01-949 6171

Detroit Diesel Allison,
PO Box 6,
London Road,
Wellingborough,
Northamptonshire NN8 2DL
tel: 0933-71122

Deutz Engines Ltd,
Riverside Road,
London SW17 0UT
tel: 01-946 9161

Eaton Ltd,
Truck Components Marketing,
Eaton House,
Staines Road,
Hounslow,
Middlesex TW4 5DX
tel: 01-572 7313

L Gardner & Sons Ltd,
Barton Hall Engine Works,
Patricroft,
Eccles,
Manchester M30 7WA
tel: 061-789 2201

GKN Axles Ltd,
Heavy Division,
Abbey Road,
Kirkstall,
Leeds LS5 3NF
tel: 0532-584611

Laycock Engineering Ltd,
Archer Road,
Millhouses,
Sheffield S8 0JY
tel: 0742-368221

Lipe Rollway Ltd,
York Avenue,
Haslingden,
Rossendale,
Lancashire BB4 4HU
tel: 0706-228321

Maxwell Brockhouse Bus Transmissions Ltd,
1 Soarbank Way,
Bishop Meadow Road,
Loughborough,
Leicestershire.
tel: 0509-213453

Perkins Engines Ltd,
Peterborough PE1 5NA
tel: 0733-67474

Rockwell CVC,
St Martin's House,
1 Hammersmith Grove,
London W6 0NX
tel: 01-741 9411

Rolls Royce Motors Ltd,
Diesel Division,
Whitchurch Road,
Shrewsbury,
Salop SY1 4DP
tel: 0743-52262

Self-Changing Gears Ltd,
Lythalls Lane,
Coventry CV6 6FY
tel: 0203-88881

Spicer Clutch Division,
Racecourse Road,
Wolverhampton,
West Midlands WV6 0NE
tel: 0902-710484

Directory of Manufacturers and Trade Associations

Spicer Drivetrain Products,
Norwich House,
45 Poplar Road,
Solihull,
West Midlands B91 3AW
tel: 021-704 4243

Universal Driveshafts Ltd,
Ringway,
Eastern Avenue,
Lichfield,
Staffordshire WS13 7SF
tel: 05432-55831

Voith Engineering Ltd,
Ambassador House,
BrigstockRoad,
Thornton Heath,
Surrey CR4 7JG
tel: 01-689 0741

ZF Gears (Great Britain) Ltd,
Abbeyfield Road,
Lenton,
Nottingham NG7 2SX
tel: 0602-869211

Fifth Wheels

Davies Magnet Works Ltd,
Thundridge,
Ware,
Hertfordshire.
tel: 0920-2287

Dayton-Walther Ltd,
Astmoor Industrial Estate,
75/76 Brindley Road,
Runcorn,
Cheshire WA7 1BR
tel: 09285-66533

George Fischer Sales Ltd,
46 Eagle Wharf Road,
London N1 7EE
tel: 01-253 1044

Jost (GB) Ltd,
Heywood Industrial Estate,
Pilsworth Road,
Heywood,
Lancashire OL10 2SQ
tel: 0706-624031

VBG,
John R Billows Group,
Truck Equipment Division,
Pytchley Lodge Road,
Kettering,
Northamptonshire NN15 6JJ
tel: 0536-516233

York Truck Equipment Ltd,
St Mark's Road,
Corby,
Northamptonshire.
tel: 06363-3561

Fuel Monitoring Systems

Ai Industrial,
London Road,
Pampisford,
Cambridge CB2 4EF
tel: 02323-834420

Centaur Electronic Systems Ltd,
16 Shaw Road,
Oldham OL1 3LQ
tel: 061-620 6438

County Pumps Ltd,
West Street,
Shutford,
Banbury,
Oxfordshire.
tel: 0295-78746

Fuel Monitoring Systems Ltd,
28/30 Grafton Road,
New Malden,
Surrey KT3 3AA
tel: 01-949 5521

Industrial Fuel Installations,
Petrochem House,
124/126 Haydons Road,
Wimbledon,
London SW19 1AE
tel: 01-542 9656

Petromatic Securi-Key Ltd,
Kelgray House,
Sandy Lane,
Crawley Down,
West Sussex RH10 4HS
tel: 0342-715066

Springfield Controls Ltd,
PO Box 71,
Bridge Works,
Stafford Road,
Wolverhampton WV10 6HS
tel: 0902-29039

E W Taylor Fuel Control Ltd,
Riverside Works,
Cambridge Road,
Harlow,
Essex CM20 2ET.
tel: 0279 38727

Timeplan Ltd,
Britannia House,
67 Old Woking Road,
West Byfleet,
Weybridge,
Surrey KT14 6LF
tel: 09323-40454

Transflo Instruments Ltd,
71/79 Loose Road,
Maidstone,
Kent ME15 7BY
tel: 0622-683888

LPG Conversion Specialists

Cam-Gas (Cambridge) Ltd,
Trafalgar Way,
Bar Hill,
Cambridge.
tel: 0954-81132

Dual-Fuel Systems Ltd,
Unit 8,
Britannia Estates,
Leagrave Road,
Luton,
Bedfordshire LU3 1RJ
tel: 0582-414090

Fleetgas Installation Systems Ltd,
(Lovato)
Italian Car Centre,
Beavor Lane,
King Street,
Hammersmith,
London W6 9BL
tel: 01-741 0772

HKL Gas Power Ltd,
(Impco),
Unit 7,
New Bartholomew Street,
Birmingham B5 5QS
tel: 021-643 7855

Renzo Landi (GB) Motor Gas
Equipment Ltd,
Bradford Road,
Wrenthorpe,
Wakefield,
West Yorkshire WF2 0QH
tel: 0924-74652

Solex Landi Hartog,
Capitol Park,
Capitol Way,
London NW9 0EW
tel: 01-205 4645

Oil Companies

BP Oil Ltd,
BP House,
Victoria Street,
London SW1E 5NJ
tel: 01-834 9090

Burmah-Castrol Ltd,
Burmah House,
Pipers Way,
Swindon,
Wiltshire SN3 1RE
tel: 0793-30151

Chevron Oil (UK) Ltd,
Rothschild House,
Whitgift Centre,
Croydon CR9 3QQ
tel: 01-686 0444

Conoco Ltd,
Berkeley Square House,
Berkeley Square,
London W1X 5PB
tel: 01-493 1235

Esso Petroleum Co Ltd,
Esso House,
Victoria Street,
London SW1E 5JW
tel: 01-834 6677

Directory of Manufacturers and Trade Associations

Gulf Oil (GB) Ltd,
6 Grosvenor Place,
London SW1
tel: 01-235 5011

Mobil Oil Company Ltd,
Mobil House,
54/60 Victoria Street,
London SW1E 6QB
tel: 01-828 9777

Shell UK Oil,
Shell Mex House,
Strand,
London WC2R 0DX
tel: 01-438 3000

Silkolene,
Dalton & Co Ltd,
Silkolene Oil Refinery,
Belper,
Derbyshire DE5 1WF
tel: 077382-4151

Texaco Ltd,
1 Knightsbridge Garden,
London SW1
tel: 01-584 5000

Power Take-Off

The Drum Engineering Co Ltd,
PO Box 178
Edward Street Works,
Tong Street,
Bradford,
West Yorkshire BD4 9SH
tel: 0274-683131

Eaton Ltd,
Eaton House,
Staines Road,
Hounslow,
Middlesex TW4 5DX
tel: 01-572 7313

Edbro International Ltd,
Lever Street,
Bolton,
Lancashire BL3 6DJ
tel: 0204-28888

Telehoist Ltd,
Manor Road,
Cheltenham,
Gloucestershire GL51 9SH
tel: 0242-21355

ZF Gears (Great Britain) Ltd,
Abbeyfield Road,
Lenton,
Nottinghamshire.
tel: 0602-869211

PSV Manufacturers and Distributors

Walter Alexander and Co
(Coachbuilders) Ltd,
91 Glasgow Road,
Falkirk FK1 4JB
tel: 0324-21672

Asco Ltd,
Blessington,
County Wicklow,
Ireland.
tel: Naas 65305

Bedford Commercial Vehicles,
PO Box 3,
Luton,
Bedfordshire LU2 0SY
tel: 0582-21122

Bombardier (Ireland) Ltd,
Shannon Industrial Estate,
County Clare,
Ireland.
tel: 061-61522

Bova
see Moseley

Caetano
see Moseley

DAF Bus,
Thames Trading Estate,
Fieldhouse Lane,
Marlow,
Buckinghamshire SL7 1LW
tel: 06284-6955

203

Carlton PSV Sales,
Doncaster Road,
Oldcotes,
Worksop,
Nottinghamshire.
tel: 0909-730249

Duple Coachbuilders Ltd,
Vicarage Lane,
Blackpool FY4 4EN
tel: 0253-62251

Duple (METSEC) Ltd,
Broadwell Works,
Oldbury,
Warley,
West Midlands B69 4HE
tel: 021-552 2929

Eastern Coach Works Ltd,
Eastern Way,
Lowestoft NR32 2HG
tel: 0502-61751

East Lancashire Coachbuilders Ltd,
Whalley New Road Works,
Blackburn BB1 6JG
tel: 0254-57061

Ensign Bus Co Ltd,
Arterial Road,
Purfleet,
Essex RM16 1TB
tel: 04026-5656

Erringtons of Evington,
Evington,
Leicestershire
tel: 0533-730421

Ford Motor Co Ltd,
Eagle Way,
Brentwood,
Essex CM13 3BW
tel: 0277-253000

Hestair Dennis Ltd,
Woodbridge Works,
Guildford,
Surrey GU2 5XP
tel: 0483-571271

HKS Coachworks,
Bowburn Industrial Estate,
Bowburn,
County Durham.
tel: 0385-770711

Jonchkheere
see Roeselare

Kassbohrer (UK) Ltd,
The Coach Centre,
Bordon,
Hampshire.
tel: 04203-4998

Leyland Bus,
Bow Lane,
Leyland,
Preston PR5 1SN
tel: 077-44-24241

Leyland National Co Ltd,
Workington,
Cumbria.
tel: 0900-4343

MAN-VW Truck and Bus Ltd,
Frankland Road,
Blagrove,
Swindon,
Wiltshire SN5 8YU
tel: 0793-40231

Marshall of Cambridge (Engineering) Ltd,
Airport Works,
Newmarket Road,
Cambridge CB5 8RX
tel: 0223-61133

Mercedes-Benz (UK) Ltd,
Mercedes-Benz Centre,
Millington Road,
Hayes,
Middlesex UB3 4DS
tel: 01-573 7777

Metro-Cammell Weymann Ltd,
PO Box 248,
Leigh Road,
Washwood Heath,
Birmingham B8 2YJ
tel: 021-327 4777

Directory of Manufacturers and Trade Associations

Moseley Group PSV Ltd,
Derby Road,
Loughborough,
Leicestershire
tel: 0509-213232

Neoplan
see Carlton

Northern Counties Motor and
Engineering Co Ltd,
Enfield Street,
Wigan,
Lancashire WN1 8DY
tel: 0942-212135

Plaxtons (GB) plc,
Coach and Bus Bodybuilding
Division,
PO Box No 2,
Castle Works,
Seamer Road,
Scarborough
Yorkshire YO12 4DQ
tel: 0723-63311

Reeve Burgess Ltd,
Bridge Street,
Pilsley,
Chesterfield S45 8HF
tel: 0773-872292

Charles H. Roe Ltd,
Cross Gates Carriage Works,
Leeds LS15 8SU
tel: 0532-645182

Roeselare Sales Ltd,
Beckett House,
14 Billing Road,
Northampton NN9 5AW
tel: 0604-31913

Scania (Great Britain) Ltd,
Tongwell,
Milton Keynes,
Buckinghamshire MK15 8HB
tel: 0908-614040

van Hool
*see Moseley Group
PSV Ltd*

van Rooijen
see HKS Coachworks

Volvo Bus (GB) Ltd,
Kilwinning Road,
Irvine,
Ayrshire.
tel: 0294-74120

Wadham Stringer (Coachbuilders)
Ltd,
Hambledon Road,
Waterlooville,
Hampshire.
tel: 07014 58211

Ward Motors Ltd,
Appleton Works,
Holmfirth Road,
Shepley,
Huddersfield,
West Yorkshire.
tel: 0484-606606

Robert Wright and Son Ltd,
Cushendall Road,
Ballymena,
Northern Ireland.
tel: Ballymena 41212

W S Yeates plc,
Bakewell Road,
Loughborough,
Leicestershire.
tel: 0509-217777

Hubbard Engineering Company Ltd,
Hillview,
Otley,
Ipswich,
Suffolk IP6 9NP
tel: 047-339522

Petter Refrigeration Ltd,
Hamble,
Southampton,
Hampshire.
tel: 0703-453631

205

Thermo King Europe,
Monivea Road,
Mervue,
Galway,
Ireland.
tel: 091-61231

Smiths Industries,
Automotive and Industrial Systems,
Fradley,
Nr Lichfield,
Staffordshire WS13 8NE
tel: 05432 24181

TransFrig Ltd,
Cranbourne Road,
Gosport,
Hampshire.
tel: 07017-88131

Road Speed Limiters

Econocruise Ltd,
180 Wood Street,
Rugby
CV21 2NP
tel: 0788-74431

Fidus Controls,
Denbigh Road,
Bletchley,
Milton Keynes,
Bucks MK1 1DH
tel: 0908-368351

Kysor Industrial (Great Britain) Ltd,
Bridge Works,
North Street,
Whitworth,
Lancs OL12 8RJ
tel: 070685-2111

Lucas Kienzle Instruments Ltd,
36 Gravelly Industrial Park,
Birmingham B24 8TA
tel: 021-328 5533

Romatic Ltd,
Unit 3,
Derwent Street Trading Estate,
Salford,
Manchester.
tel: 061-834 0727

Improve profitability by saving fuel and reducing the risk of expensive breakdowns.

KYSOR manufacture a complete range of vehicle protection systems designed to save fuel... extend engine life and protect against unscheduled breakdowns.

KYSOR protection comes cheap... the alternative could prove extremely expensive so why not get the facts on KYSOR protection now.

● **Radiator Shutters that save you up to 15% on your fuel bills.**
● **Engine Protection systems that protect owner and driver alike.**
● **Road Speed control systems that put you in control of your fuel costs.**
● **Air Conditioners to improve driver comfort and efficiency.**

Please ask for literature on Automatic Radiator Shutters, Engine Protection Systems or Kymax Maximum Speed Control.

One day you'll say thanks to KYSOR

Kysor Industrial
(Great Britain) Ltd.,
Bridge Works,
North Street,
Whitworth,
Lancs. OL12 8RJ,
England.
Tel: (070 685) 2111.
Telex: 63264

International Masters
of Temperature Control

Directory of Manufacturers and Trade Associations

Smiths Industries,
50 Oxgate Lane,
Cricklewood,
London NW2 7JB
tel: 01-452 3333

Safety Devices

Aerodyne Equipment,
30 Dudley Gardens,
Harrow,
Middlesex HA2 0DQ
tel: 01-422 1705
('wetmaster' spray suppression mudflap)

Berm Optical Products Ltd,
Barmby Moor,
York YO4 5RJ
tel: 07592-2161
(magnifying rear view lenses)

Hope Technical Developments Ltd,
High Street,
Ascot,
Berkshire SL5 7HP
tel: 0990-24855
(Hope 'safe-T-bar' rear guard and Hope anti-jack-knife device)

Intertruck Heavy Duty Division,
8 Rothersthorpe Crescent,
Northampton NN4 9JQ
tel: 0604-66233
('Raintrapper' spray guard)

Monsanto plc,
Thames Tower,
Burleys Way,
Leicester LE1 3TP
tel: 0533-27153
('Clear Pass' spray guard)

Quinton Hazell plc,
Hazell House,
Leamington Spa,
Warwickshire CV32 6RF
tel: 0926-29121
(QH 'Underider' rear guard)

Schlegel (UK) Engineering Ltd,
Henlow Industrial Estate,
Henlow Camp,
Bedfordshire.
tel: 0462-812812
(spray guard)

TI Tube Products Ltd,
PO Box 13,
Popes Lane,
Oldbury,
Warley,
West Midlands B69 4PE
tel: 021-552 1511
(TI 'Rearguard' rear guard)

YEC Audible Warning Systems,
Brigade Works,
Brigade Street,
Blackheath Village,
London SE3 0TW
tel: 01-852 3261
(audible reversing alarms)

Sleeper Cab Conversions

Bonwitco,
38 Ebrington Street,
Kingsbridge,
Devon TQ7 1DE
tel: 0548-2453

Dorm-O-Cab Ltd,
Lion Barn Industrial Estate,
Needham Market,
Ipswich,
Suffolk IP6 8NZ
tel: 0499-721811

Hatcher Components Ltd,
Broadwater Road,
Framlingham,
Suffolk.
tel: 0728-723675

Jennings Coachworks Ltd,
Second Avenue,
Weston Road,
Crewe,
Cheshire CW1 1DJ
tel: 0270-583358

Locomotors Ltd,
Barlows Lane,
Andover,
Hampshire SP10 2HD
tel: 0264-51221

Unity Conversions,
Rigby Street,
Great Lever,
Bolton,
Lancashire BL3 6PB
tel: 0204-26088

Suspension Systems

JWE Banks and Sons Ltd,
Crowland,
Peterborough PE8 0JP
tel: 0733 210316

Dunlop Ltd,
Engineering Group,
Suspensions Division,
Holbrook Lane,
Coventry CV6 4AA
tel: 0203-88733

Granning Suspensions UK,
42A Longbridge Road,
Barking,
Essex.
tel: 01-594 1771

Hands Neway Ltd,
Hucclecote,
Gloucester GL3 4AD
tel: 0452-69321

Hendrickson (Europe) Ltd,
Courtsdown,
Little Island,
Cork,
Eire.
tel: 021 953211

Norde Suspensions Ltd,
Sywell Airport,
Sywell,
Northamptonshire NN6 0BU
tel: 0604-493161

Primrose Third Axle Co Ltd,
Lever Works,
Slater Street,
Ewood,
Blackburn,
Lancashire.
tel: 0254-56031

Rubery Owen-Rockwell Ltd,
Parts and Accessories Division,
1/3 Cranford Court,
Hardwick Grange,
Woolston,
Warrington,
Cheshire.
tel: 0925-823023

York Technical Services Ltd,
Rockingham Road Industrial Estate,
Market Harborough,
Leicestershire.
tel: 0858-65431

Tail Lifts

BCT Equipment,
Adam House,
58/66 Birmingham Road,
Kidderminster,
Worcestershire DY10 2SH
tel: 0562 740555

Primrose Tail Lifts,
Lever Works,
Slater Street,
Ewood,
Blackburn,
Lancashire.
tel: 0254-56031

PTH (Clitheroe) Ltd,
Woone Lane,
Clitheroe,
Lancashire BB7 1BW
tel: 0200-23784

Ratcliff (Tail Lifts) Ltd,
Bessemer Road,
Welwyn Garden City,
Hertfordshire AL7 1ET
tel: 07073-25571

Directory of Manufacturers and Trade Associations

Scan Lift,
Ray Smith Demountables Ltd,
Botolph Bridge,
Oundle Road,
Peterborough.
tel: 0733-63936

Tellift,
Telehoist Ltd,
Manor Road,
Cheltenham,
Gloucestershire GL51 9SH
tel: 0242 21355

Zepro Tails Lifts,
2 Aintree Drive,
Leamington Spa,
Warwickshire CY32 7TU
tel: 0926-30966

38 Tonne Conversion Specialists

Crane Fruehauf Ltd,
Toftwood,
Dereham,
Norfolk,
tel: 0362-5353

Dean's Engineering Services,
15 Cranmer Road,
The Meadows Industrial Estate,
Derby DE2 6JL
tel: 0332-45621

Granning Suspensions UK,
42A Longbridge Road,
Barking,
Essex.
tel: 01-591 2786

Lyka Ltd,
382 Blackpool Road,
Preston,
Lancashire PR2 2DS
tel: 0772-727927

Murphy Transport Services,
852 Melton Road,
Thurmaston,
Leicester.
tel: 0533-69611

PTH (Clitheroe) Ltd,
Primrose Road,
Clitheroe,
Lancashire.
tel: 0200-23784

Southworth Mechanical,
Rawlinson Lane,
Heath Charnock,
Chorley,
Lancashire PR6 9JX
tel: 0257-481607

York Truck Equipment Ltd,
Corby,
Northamptonshire NN18 8AH
tel: 05363-3561

Tipping Gear

Allison Harsh,
Allisons Plant Ltd,
Manor Buildings,
The Airfield,
Pocklington,
York YO4 2HB
tel: 07592-2001

Anthony Carrismore Ltd,
Harelaw Industrial Estate,
Stanley,
County Durham.
tel: 0207-32461

Edbro Ltd,
Lever Street,
Bolton,
Lancashire BL3 6DJ
tel: 0204-28888

Hyva (UK) Ltd,
Taylor Industrial Estate,
Warrington Road,
Risley-Warrington,
Cheshire WA3 6BH
tel: 0925-764146

Multilift Ltd,
Ainsdale Drive,
Harlescott Industrial Estate,
Shrewsbury,
Salop SY1 3TJ
tel: 0743-58009

Telehoist Ltd,
Manor Road,
Cheltenham,
Gloucestershire GL51 9SH
tel: 0242-21355

Trade and Professional Associations

Association of British Chamber of Commerce,
68 Queen Street,
London EC4N 1SN
tel: 01-248 7211

Association of Vehicle Recovery Operators,
6th Floor,
Epic House,
Charles Street,
Leicester LE1 3SH
tel: 0533-538915

Automobile Association,
Fanum House,
Basingstoke,
Hampshire RG21 2EA
tel: 0256-20123

British Association of Removers,
279 Gray's Inn Road,
London WC1X 8SY
tel: 01-837 3088

British Road Federation,
388/396 Oxford Street,
London W1N 9HE
tel: 01-499 0281

British Standards Institution,
2 Park Street,
London W1A 2BS
tel: 01-629 9000

British Vehicle Rental Association,
11 North Pallant,
Chichester,
West Sussex.
tel: 0243-786782

Bus & Coach Council,
Sardinia House,
52 Lincoln's Inn Fields,
London WC2A 3LZ
tel: 01-831 7546

Centre for Physical Distribution Management,
Management House,
Cottingham Road,
Corby,
Northamptonshire.
tel: 05366-42222

Chartered Institute of Transport,
80 Portland Place,
London W1N 4DP
tel: 01-580 5216

Continental Freight Drivers Club,
354 Fulham Road,
London SW10
tel: 01-351 3522

The Electricity Council,
30 Millbank,
London SW1P 4RD
tel: 01-834 2333

Electric Vehicle Association,
30 Millbank,
London SW1P 4RD
tel: 01-834 2333

Electric Vehicle Development Group,
59 Colebrooke Row,
London N1
tel: 01-359 7352

European Conference of British Bus & Coach Operators (ECBO),
Sardinia House,
52 Lincoln's Inn Fields,
London WC2A 3LZ
tel: 01-831 7546

Directory of Manufacturers and Trade Associations

Fire Fighting Vehicle Manufacturers Association,
c/o SMMT,
Forbes House,
Halkin Street,
London SW1X 7DS
tel: 01-235 7000

Freight Transport Association,
Hermes House,
St John's Road,
Tunbridge Wells,
Kent TN4 9UZ
tel: 0892-26171

Garage Equipment Association,
Forbes House,
Halkin Street,
London SW1X 7DS
tel: 01-235 7111/2

Institute of Automotive Engineer Assessors,
1 Love Lane,
London EC2V 7JJ
tel: 01-606 8744

Institute of British Carriage & Automobile Manufacturers,
'Thames Meadow',
Henley Road,
Shillingford,
Oxford OX9 8EZ
tel: 086732-8263

Institute of Materials Handling,
St Ives House,
St Ives Road,
Maidenhead,
Berkshire.
tel: 0628-28011

Institute of Motor Industry,
Fanshaws,
Brickendon,
Hertford.
tel: 099-286282/3

Institute of Road Transport Engineers,
1 Cromwell Place,
Kensington,
London SW7 2JF
tel: 01-589 3744

Institute of Trading Standards Administration,
Estate House,
319D London Road,
Hadleigh,
Benfleet,
Essex SS7 2BN
tel: 0702 558179

Institute of Transport Administration,
32 Palmerston Road,
Southampton SO1 1LL
tel: 0703-31380

Institute of Transport Managers,
Pengo House,
131 Duckmoor House,
Bristol BS3 2BJ
tel: 0272-633869

Institution of Body Engineers,
'Thames Meadow',
Henley Road,
Shillingford,
Oxford OX9 8EZ
tel: 086732-8263

Institution of Mechanical Engineers,
1 Birdcage Walk,
London SW1
tel: 01-222 7899

Motor Agents' Association,
201 Great Portland Street,
London W1N 6AB
tel: 01-580 9122

Motor Factors' Association,
16A The Broadway,
London SW19 1RF
tel: 01-946 3499

211

Motor Industry Research
Association,
Watling Street,
Nuneaton,
Warwickshire CV10
tel: 0682-348541

National Tyre Distributors
Association,
Broadway House,
The Broadway,
London SW19
tel: 01-540 3859

Road Haulage Association,
104 New Kings Road,
Fulham,
London SW6 4LN
tel: 01-736 1183

Royal Automobile Club,
83/85 Pall Mall,
London SW1Y 5HW
tel: 01-839 7050

Society of Motor Manufacturers &
Traders,
Forbes House,
Halkin Street,
London SW1X 7DS
tel: 01-235 7000

Transport Association,
9th Floor,
Centre City Tower,
7 Hill Street,
Birmingham B5 4UU
tel: 021-643 5494

Transport & Road Research
Laboratory,
Crowthorne,
Berkshire.
tel: 03446-31313

Transport Trust,
Terminal House,
Shepperton,
Middlesex TW17 8AS
tel: 09322-28950

**Trailer Manufacturers/
Bodybuilders**

Acoma (Bilston) Ltd,
Dale Street,
Bilston,
West Midlands WV14 7JX
tel: 0902-42331

Alcan Transport Products Ltd,
Hurricane Way,
Norwich Airport Industrial Estate,
Norwich NR6 6HE
tel: 0603-49241

Jack Allen (Motor Bodies),
Municipal House,
Buckingham House,
Buckingham Street,
Birmingham B19 3HS
tel: 021-236 7124

Arlington Bodybuilders Ltd,
High Road,
Ponders End,
Enfield,
Middlesex EN3 4EG
tel: 01-804 1266

Atkinson & Co (Clitheroe) Ltd,
Kendal Street,
Clitheroe,
Lancashire BB7 1NZ
tel: 0200-22211

Bedwas Bodyworks Ltd,
Pant Glas Industrial Estate,
Bedwas,
Newport,
Gwent NP1 9XA
tel: 0222-885781

Besco Bodies (Northampton) Ltd,
Ross Road,
Weedon Road Industrial Estate,
Northampton NN5 5AX
tel: 0604-581386

Boalloy Ltd,
Radnor Park Estate,
West Heath,
Congleton,
Cheshire CW12 4QA
tel: 02602-5151

Bostock and Barsby Ltd,
Hazel Way,
Barwell,
Leicestershire.
tel: 0455 43303

Brade-Leigh Products Ltd,
Albion Industrial Estate,
Oldbury Road,
West Bromwich,
West Midlands B70 9EH
tel: 021-553 4361

Broshuis Trailers Ltd,
PO Box 18,
Princes Risborough,
Aylesbury,
Buckinghamshire HP17 9TG
tel: 084474-3582

Buckingham Vehicles Ltd,
Dalehouse Lane,
Kenilworth,
Warwickshire CV8 1LE
tel: 0926-512411

Buckstones Motor Bodies Ltd,
Trent Mill Industrial Estate,
Duchess Street,
Shaw OL2 7UT
tel: 0706-842551

Bulkmobile Ltd,
Botolph Bridge,
Peterborough.
tel: 0733-313151

Burton and Greaves (Batley) Ltd.,
Victoria Avenue,
Batley,
West Yorkshire.
tel: 0924-472366

Carmichael Fire and Bulk Ltd,
Gregory's Mill Street,
Worcester,
Worcestershire WR3 8BE
tel: 0905-21381

Anthony Carrimore (Sales) Ltd,
North Road,
Harelaw,
Stanley,
County Durham DH9 8HJ
tel: 0207-32461

Carryfast Ltd,
Customline Manufacturing Division,
Mill Street East,
Dewsbury,
West Yorkshire WF12 9AP
tel: 0924-463184

Carrymaster Ltd.,
Carrymaster House,
Askern Road,
Carcroft,
Doncaster.
tel: 0302 723411

Cartwright Freight Systems Ltd,
Atlantic Street,
Broadheath,
Altrincham,
Cheshire.
tel: 061-941 4023

CMC Bodies,
Blenheim Way,
Market Deeping,
Lincolnshire.
tel: 0778-344291

Coachwork Conversions Ltd,
Norfolk Street,
Colne,
Lancashire BB8 9JW
tel: 0282-866724

Collis Gold Containers Ltd,
Fort Wallington,
Fareham,
Hampshire.
tel: 0329-232595

213

Concargo Ltd,
Winterstoke Road,
Weston-super-Mare,
Avon BS24 9AH
tel: 0934-28221

Ben Cooper Engineering Ltd,
Bridge Motor Works,
Claydon,
Ipswich,
Suffolk IP6 0HU
tel: 0473-830444

Crane Fruehauf Ltd,
Toftwood,
Dereham,
Norfolk.
tel: 0362-5353

Craven Tasker Ltd,
Joe Lane,
Catterall,
Garstang,
Lancashire PR3 0QD
tel: 09952 4211

Laurence David Ltd,
St Peters House,
1 Bishops Road,
Peterborough PE1 1YE
tel: 0733 47380

Don-Bur (bodies and trailers) Ltd,
Mossfield Road,
Adderley Green,
Longton,
Stoke-on-Trent,
Staffordshire ST3 5BW
tel: 0782-329329

Dormobile Ltd,
Tile Kiln Lane,
Folkestone,
Kent CT19 4PD
tel: 0303-76321

Dunnspencer Bulkflo Ltd,
Ilton Works,
Ilton,
Ilminster,
Somerset TA19 9DU
tel: 04605-2931

R A Dyson (1981) Ltd,
Bedfordia House,
76/80 Grafton Street,
Liverpool L8 6RH
tel: 051-708 8440

Edbro International Ltd,
Lever Street,
Bolton,
Lancashire BL3 6DJ
tel: 0204-28888

Fergusons Tankers Ltd,
Limberline Road,
Hilsea,
Portsmouth,
Hampshire PO3 5JG
tel: 0705-64231

Fulton and Wylie Ltd,
East Road,
Irvine,
Ayrshire.
tel: 0294-72929

J Gardiner and Sons (Edenbridge) Ltd,
Littlebrowns Works,
Edenbridge,
Kent TN8 6LH
tel: 0732-862335

Gloster Saro Ltd,
Hucclecote,
Gloucester.
tel: 0452-69321

Glover, Webb & Liversidge Ltd,
Hamble Lane,
Hamble,
Hampshire SO3 5NY
tel: 0421-222811

Gray & Adams Ltd,
South Road,
Fraserburgh,
Aberdeenshire AB4 5HU
tel: 03462-200013

Hawson-Garner Ltd,
Brooklands Close,
Windmill Road,
Sunbury-on-thames,
Middlesex.

HCB-Angus Ltd,
South Hampshire Industrial Park,
Testwood,
Southampton SO4 35A
tel: 0703-865766

Hestair Eagle Ltd,
The Saltisford,
Warwick,
Warwickshire CV34 5XW
tel: 0926-44321

Ingimex Ltd,
Halesfield 9,
Telford,
Shropshire.
tel: 0952-585833

Jennings Coachworks Ltd,
Second Avenue,
Weston Road,
Crewe,
Cheshire.
tel: 0270-583358

Kalmar Kockum Ltd,
Boyn Valley Industrial Estate,
Maidenhead,
Berkshire.
tel: 0628-39944

King Trailers Ltd,
Riverside,
Market Harborough,
Leicestershire LE16 7PX
tel: 0858-67361

Kurtrans Developments Ltd,
Unit D1,
Halesfield 9,
Telford,
Shropshire.
tel: 0957-582744

David Macrill Engineering Ltd,
Robert Body Works,
Bury St Edmunds,
Suffolk IP33 3PH
tel: 0284-67444

Mackworth Bodyworks Ltd,
Ashby Road Central,
Shepshed,
Loughborough
Leicestershire LE12 5BR
tel: 05095-2121

M & G Trailers (Lye) Ltd,
Hayes Lane,
Lye,
Stourbridge,
West Midlands DY9 8PA
tel: 0348482-5221

Marshall of Cambridge
(Engineering) Ltd,
Airport Works,
Cambridge CB5 8RX
tel: 0223-61133

Massey (International
Coachbuilders) Ltd,
High Street,
Market Weighton,
York YO4 3AD
tel: 0696-72361

Merriworth (Engineering) Ltd,
Richmer Road,
Erith,
Kent.
tel: 03224-33441

Metalair Ltd,
Sutton Bridge,
Lincolnshire PE12 9XE
tel: 0406-350000

Multilift Ltd,
Ainsdale Drive,
Harlescott Industrial Estate,
Shrewsbury.
tel: 0743-68009

215

Neville Charrold Ltd,
Bradder Works,
Mansfield,
Nottinghamshire NG18 5DQ
tel: 0623-25354

George Neville Truck Equipment Ltd,
Lindleys Lane,
Kirkby-in-Ashfield,
Nottinghamshire.
tel: 0623-752601

Nooteboom UK,
42A Longbridge Road,
Barking,
Essex.
tel: 01-594 1771

Norwich Coachworks Ltd,
Burton Road,
Norwich.
tel: 0603-49221

PEM Trailers Ltd,
Wallisdown Industrial Estate,
Wallisdown Road,
Bournemouth,
Dorset BH11 8PT
tel: 0202-512233

AC Penman Ltd,
Heathall,
Dumfries,
Dumfriesshire DG1 3NY
tel: 0387-2784

Pilcher-Greene Ltd,
Victoria Gardens,
Burgess Hill,
West Sussex RH15 9NA
tel: 04446-5707

Powell Duffryn Engineering Ltd,
Llantrisant,
Pontyclun,
Mid Glamorgan.
tel: 0443-222301

Primrose Third Axle Co Ltd,
Ewood,
Blackburn,
Lancashire.
tel: 0254-56031

Redment Engineering Ltd,
Springwood Industrial Estate,
Rayne Road,
Braintree,
Essex.
tel: 0376-21900

Reptrail Ltd,
Claycliffe Road,
Barugh Green,
Barnsley,
South Yorkshire.
tel: 0226-292705

Reynolds Boughton Ltd,
Bell Lane,
Amersham,
Buckinghamshire.
tel: 02404-4411

Charles Roberts Engineering Ltd,
Horbury Junction,
Wakefield,
West Yorkshire WF4 5ES
tel: 0924-274681

Rootes Maidstone Ltd,
Mill Street,
Maidstone,
Kent ME15 6YD
tel: 0622-53333

Ryland Tankers Ltd,
Saltwells Road,
Netherton,
Dudley,
West Midlands.
tel: 0384-60181

Seadyke Freight Systems Ltd,
Nene Parade,
Wisbech,
Cambridgeshire.
tel: 0945-65311

Silverdale Bodies Ltd,
Station Road,
Coleshill,
Birmingham.
tel: 0675-65414

Directory of Manufacturers and Trade Associations

Southfields Coachworks Ltd,
Bakewell Road,
Loughborough,
Leicestershire.
tel: 0509-66461

Sparshatts (Hants) Ltd,
London Road,
Hilsea,
Portsmouth,
Hampshire.
tel: 0705-693426

Spenborough Engineering Co Ltd,
Union Road,
Heckmondwyke,
West Yorkshire WF15 7JY
tel: 0924-403411

Spencer, Abbott (Engineers) Ltd,
51 Tyburn Road,
Birmingham B24 8NN
tel: 021-327 2711

Star Bodies (NFC) Ltd,
Star Works,
Spencer Street,
Hollinwood,
Oldham OL9 7JE
tel: 061-624 7938

Stevecastle Ltd,
Gas Works,
East Road,
Oundle,
Peterborough.
tel: 0515-481183

Structure-Flex Ltd,
Grave Lane,
Holt,
Norfolk NR25 6EG
tel: 026371-2911

Tamplin Engineering Ltd,
Birdham,
Chichester,
Sussex PO20 7BU
tel: 0243-512599

Tarvin Trailers Ltd,
Brook Works,
Tarvin,
Cheshire.
tel: 0829-40071

Telehoist Ltd,
Manor Road,
Cheltenham,
Gloucestershire GL51 9SH
tel: 0242-21355

Thompson Tankers,
Great Bridge Road,
Bilston,
West Midlands.
tel: 0902 43141

Tidd Strongbox Ltd,
Marston Road,
Cromwell Road Industrial Estate,
St Neots,
Cambridgeshire PE19 2HD
tel: 0480-214500

Tilt Techniek Ltd,
4 Atcost Road,
Off River Road,
Barking,
Essex.
tel: 01-594 7634

Tipmaster Ltd,
Rigg Approach,
Lea Bridge Road,
Leyton.
tel: 01-539 0611

Transquip International Trailers Ltd,
15 Princewood Road,
Earlstree Industrial Estate North,
Corby,
Northamptonshire NN17 2AP
tel: 05366-67555

Universal Bulk Handling Equipment Ltd,
Orrell Lane,
Burscough,
Lancashire L40 0SL
tel: 0704-892611

217

Vaile & Co (Dorset) Ltd,
Churchfoot Lane,
Hazelbury Bryan,
Dorset.
tel: 02586-225

Vanplan Ltd,
Chesford Grange,
Grange Estate,
Warrington WA1 4RA
tel: 0925-821512

Wadham Stringer (Coachbuilders) Ltd,
Hambledon Road,
Waterlooville,
Hampshire.
tel: 07014-2661

B Walker & Son,
Gammons Lane,
Watford,
Hertfordshire.
tel: 0923-25816

Weighlifter Bodies Ltd,
Broughton Brigg,
South Humberside DN20 0LA
tel: 0652-56331

Welford Truck Bodies Ltd,
Hainge Road,
Tividale,
Warley,
West Midlands.
tel: 021-557 2721

Welmech (Staffs) Engineering Co Ltd,
Saxon Works,
Royal Street,
Fenton,
Stoke-on-Trent,
Staffordshire.
tel: 0782-314313

Whale Tankers Ltd,
Ravenshaw Lane,
Solihull,
West Midlands B91 2SU
tel: 021-704 3191

EM Wilcox Ltd,
Royce Road,
Carr Road Industrial Estate,
Peterborough,
Cambridgeshire PE1 5YB
tel: 0733-68585

Wilsdon & Co Ltd,
Industrial Estate,
Lode Lane,
Solihull,
West Midlands B91 2JR
tel: 021-705 1177

Wincanton Bodywork,
Wincanton Garages (Westbury) Ltd,
The Ham,
Westbury,
Wiltshire BA13 4HB
tel: 0373-823534

Windfoil,
Rash's Green Industrial Estate,
East Dereham,
Norfolk NR19 1JG
tel: 0362-67115

K & J Withey Ltd,
Gellideg Lane,
Maesycwmmer,
Hengoed,
Mid Glamorgan.
tel: 0443-812097

York Bodyline,
Hawleys Lane,
Warrington,
Cheshire WA2 8JP
tel: 0925-34111

York Trailer Co Ltd,
St Marks Road,
Corby,
Northamptonshire NN18 8AH
tel: 05366-3561

Zeromobile Ltd,
7 Mount Ephraim Road,
Tunbridge Wells,
Kent.
tel: 0892-34713

Directory of Manufacturers and Trade Associations

Turbocharger Manufacturers and Service Centres

Garrett AiResearch Ltd,
Turbocharger Division,
East Pimbo,
Skelmersdale,
Lancashire WN8 9PH
tel: 0695-22391

Holset Engineering Company Ltd,
PO Box A9,
Turnbridge,
Huddersfield,
West Yorkshire HD1 6RD
tel: 0484-22244

Roto-Master Ltd,
Rashcliffe Mills,
Albert Street,
Huddersfield,
West Yorkshire HD1 3RP
tel: 0484-512222

Schwitzer Wallace Murray Ltd,
Roydsdale Way,
Euroway Industrial Estate,
Bradford,
West Yorkshire BD4 6SE
tel: 0274-684915

Turbo Exchange,
CAV Parts & Service,
Lucas CAV Ltd,
PO Box 36,
Warple Way,
London W3 7SS
tel: 01-743 3111

Turbo International Ltd,
PO Box 19,
Dowding Road,
Lincoln LN3 4PJ
tel: 0522-38811

Tyres

The Bandag Tyre Company,
Lamberhead Industrial Estate,
Leopold Street,
Pemberton,
Wigan,
Lancashire WN5 8DH
tel: 0942-214827

Bridgestone Tyres,
Old Walsall Road,
Great Barr,
Birmingham B42 1EA
tel: 021-358 5921

John Bull Tyres,
Factory Street,
Chesterfield,
Derbyshire.
tel: 0246-77241

Continental Tyre & Rubber Co Ltd,
Ullswater Crescent,
Coulsdon,
Surrey.
tel: 01-668 2372

Dunlop Ltd,
Tyre Division,
Wood Lane,
Erdington,
Birmingham B24 9QT
tel: 021-373 2121

The Goodyear Tyre & Rubber Co
(GB) Ltd,
Stafford Road,
Bushbury,
Wolverhampton WV10 6DH
tel: 0902-22321

Michelin,
81 Fulham Road,
London SW3 6RD
tel: 01-589 1460

Pirelli Ltd,
Derby Road,
Burton-on-Trent,
Staffordshire DE18 0BH
tel: 0283-66301

Pneumant Sterling – Interscope Ltd
Midford Place
110/113 Tottenham Court Road,
London W1P 9HG
tel: 01-387 0771

Toyo Tyre (UK),
Euro House,
1394 High Road,
Whetstone,
London N20 9BH
tel: Freephone 2543

Vacu-Lug Traction Tyres Ltd,
Spitalgate Mill,
Bridge End Road,
Grantham,
Lincolnshire NG31 7HY
tel: 0476 2424

Vehicle Finance and Leasing

Arlington Motor Finance Ltd,
4 Vigo Street,
London W1
tel: 01-743 7040

Bowmaker Financial Services,
Bowmaker Ltd,
Bowmaker House,
Christchurch Road,
Bournemouth BH1 3LG
tel: 0202-22077

British Credit Trust plc,
26 High Street,
Slough SL1 1ED
tel: 0753-73211

Chartered Trust plc,
24/26 Newport Road,
Cardiff CF2 1SR
tel: 0222-484484

Forward Trust Ltd,
12 Calthorpe Road,
Edgbaston,
Birmingham B15 1QZ
tel: 021-454 6141

Lloyds & Scottish Finance Group,
Finance House,
Orchard Brae,
Edinburgh EH4 1PF
tel: 031-332 2451

Lombard North Central plc,
Lombard House,
Curzon Street,
London W1A 1EU
tel: 01-409 3434

Mercantile Credit Co Ltd,
Elizabethan House,
Great Queen Street,
London WC2B 5DP
tel: 01-242 1234

North West Securities Ltd,
North West House,
City Road,
Chester CH1 3AN
tel: 0244-315351

United Dominions Trust Ltd,
51 Eastcheap,
London EC3P 3BU
tel: 01-623 3020

Vehicle Manufacturers

Austin Rover Group Ltd,
Canley Road,
Canley,
Coventry CV5 6QX
tel: 0203-70111

Bedford Commercial Vehicles,
PO Box No 3,
Luton,
Bedfordshire LU2 0SY
tel: 0582-21122

DAF Trucks (GB) Ltd,
Thames Trading Estate,
Fieldhouse Lane,
Marlow,
Buckinghamshire SL7 1LW
tel: 06284-6955

Directory of Manufacturers and Trade Associations

Daihatsu UK Ltd,
PO Box 5,
Poulton Close,
Dover,
Kent.
tel: 0304-213030

Datsun (UK) Ltd,
see Nissan

Dodge
see Karrier

Ebro UK Ltd,
Columbia Drive,
Durrington,
Worthing,
Sussex.
tel: 0903-692299

ERF Ltd,
Sun Works,
Sandbach,
Cheshire CW11 9DN
tel: 09367-3223

Fiat Auto (UK) Ltd,
Great West Road,
Brentford,
Middlesex TW8 9DJ
tel: 01-568 8822

Fiat Trucks
see IVECO

Foden Trucks,
Sandbach,
Cheshire CW11 9HZ
tel: 09367-3244

Ford Motor Company Ltd,
Eagle Way,
Warley,
Brentwood,
Essex.
tel: 0277-253000

Freight Rover,
Tyburn Road,
Erdington Road,
Birmingham B24 8HJ
tel: 021-328 1777

Hestair Dennis Ltd,
Woodbridge Works,
Guildford,
Surrey GU2 5XP
tel: 0483-71271

Hino
HCV Motor Vehicle Distributors Ltd,
HCV House,
10 Chesford Grange,
Woolston,
Warrington,
Cheshire WA1 4RQ
tel: 0925-824026

Honda (UK) Ltd,
4 Power Road,
Chiswick,
London W4 5YT
tel: 01-995 9381

Hyundai
see International Motors Ltd

International Motors Ltd,
Ryder Street,
West Bromwich,
West Midlands B70 0EJ
tel: 021-557 9951

IVECO (UK) Ltd,
Road One,
Industrial Estate,
Winsford,
Cheshire CW7 3Q'p
tel: 06065-4950/3400

Karrier Motors Ltd,
Boscombe Road,
Dunstable,
Bedfordshire LU5 4LX
tel: 0582-64211

Kenworth,
Dando's (Motor Services) Ltd,
Chipping Sodbury,
Bristol.
tel: 0454-318311

Leyland Vehicles Ltd,
Lancaster House,
Leyland,
Preston,
Lancashire PR5 1SN
tel: 07744-21400

Magirus
see *IVECO*

MAN-VW Truck & Bus Ltd,
Frankland Road,
Blagrove,
Swindon,
Wiltshire SN5 8YU
tel: 0793-40231

Mazda (UK) Ltd,
Longfield Road,
Tunbridge Wells,
Kent TN2 3EY
tel: 0892-40123

Mercedes-Benz (UK) Ltd,
Mercedes-Benz Centre,
Millington Road,
Hayes,
Middlesex UB3 4S
tel: 01-573 7777

Mitubishi Truck & Bus Division,
Colt Car Company Ltd,
Watermoor,
Cirencester,
Gloucestershire.
tel: 0285-5777

Nissan (UK) Ltd,
Nissan House,
Columbia Drive,
Worthing,
Sussex.
tel: 0903-68561

Peugeot
see *Talbot*

Renault Trucks
see *Karrier*

Renault UK Ltd,
Western Avenue,
London W3 0RZ
tel: 01-992 3481

Reynolds Boughton Chassis Ltd,
Winkleigh Airfield,
Winkleigh,
Devon.
tel: 083783-555

Sandbach Engineering Company,
Sandbach,
Cheshire CW11 9HZ
tel: 09367-3244

Scammell Motors,
Leyland Vehicles Ltd – Heavy
Vehicle Division,
Tolpits Lane,
Watford WD1 8QB
tel: 0923-44211

Scania (Great Britain) Ltd,
Tongwell,
Milton Keynes,
Buckinghamshire MK15 8HB
tel: 0908-614040

Seddon Atkinson Vehicles Ltd,
Woodstock Factory,
Oldham,
Lancashire OL2 6HP
tel: 061-624 0566

Shelvoke & Drewry Ltd,
Icknield Way,
Letchworth,
Hertfordshire SG6 1EN
tel: 04626-6555

Subaru
see *International Motors Ltd*

Suzuki GB Ltd
46/64 Gatwick Road,
Crawley,
West Sussex RH10 2XF
tel: 0293-518000

Directory of Manufacturers and Trade Associations

Terberg,
Frank Tinsdale Ltd,
Commerce House,
Dene Drive,
Winsford,
Cheshire CW7 1AY
tel: 06065-57631

Toyota (GB) Ltd,
The Quadrangle,
Redhill,
Surrey RH1 1PX
tel: 0737-68585

VAG (United Kingdom) Ltd,
Yeomans Drive,
Blakelands,
Milton Keynes,
Bedfordshire MK14 5AN
tel: 0908-679121

Volkswagen
see VAG

Volvo Trucks (GB) Ltd,
Kilwinning Road,
Irvine,
Ayrshire.
tel: 0294-74120

White Truck Concessionaires Ltd,
Station Road,
Coleshill,
Birmingham B46 1HD
tel: 0675-64757

Washing Equipment Manufacturers

Byrange Ltd,
Albion Mill,
Helmshore Road,
Helmshore,
Rossendale,
Lancashire BB4 4JR
tel: 0706-214241

DIMA,
Anglia Distribution Ltd,
Swilland,
Ipswich,
Suffolk.
tel: 047385-327

Diversey Ltd,
Transport Industries Divisin,
Weston Favell Centre,
Northampton NN3 4PD
tel: 0604-405311

Esquire Kleindienst & Co Ltd,
Dane Works,
Great West Road,
Brentford,
Middlesex.
tel: 01-568 7751

Karcher (UK) Ltd,
Karcher House,
Beaumont Road,
Banbury,
Oxfordshire OX16 8TS
tel: 0295-67511

KEW Sales Ltd,
Carleton,
Penrith,
Cumbria.
tel: 0768-65777

Nielsen Chemicals Comercio Ltd,
Hadenham Road,
South Lowestoft Industrial Estate,
Lowestoft,
Suffolk.
tel: 0502-85637

Sellarc,
W Bateman & Company,
Garstang Road,
Barton,
near Preston,
Lancashire.
tel: 0772-862948

Smith Bros & Webb Ltd,
Britannia Works,
Arden Forest Estate,
Alcester,
Warwickshire B49 6EX
tel: 0789-763222

SP Services (Pressure Washers) Ltd,
263 Whitehall Road,
Leeds LS12 6ER
tel: 0532-791767

Warwick Pump & Engineering Co Ltd,
Berinsfield,
Oxford OX9 8LZ
tel: 0865-340322

Wesley Pressure Washers,
Wesley House,
Wortley Moor Lane,
Leeds LS12 4HT
tel: 0532-790015

Wind Deflectors

Aerodynamics Equipment,
30 Dudley Gardens,
Harrow,
Middlesex.
tel: 01-422 1705

Airshield Ltd,
(Roscoe Howard Tickle)
New Thomas Street,
Salford M6 6WP
tel: 061-736 1334

Airvane (GB) Ltd,
Anchor and Hope Lane,
Charlton,
London SE7
tel: 01-858 3781

Econofin,
Environmental Dynamics Ltd,
56 Bridge St East,
Welwyn Garden City,
Hertfordshire.
tel: 07073-30633

Europa Spoiler Ltd,
Department CM,
Stone Drive,
Colwall,
Malvern,
Worcestershire.
tel: 0684-40800

Hatcher Components Ltd,
Broadwater Road,
Framlingham,
Suffolk.
tel: 0728-723675

Sleaford Engineering Co Ltd,
Southstack,
Fulshaw Park South,
Wilmslow,
Cheshire SK19 1QF
tel: 0625-525181

Smiths Industries Ltd,
Automotive and Industrial Systems,
Fradley,
Nr Lichfield,
Staffordshire WS13 8NE
tel: 05432-24181

Strathcarron Air Management Ltd,
1 Clifton Road,
Southall,
Middlesex.
tel: 01-571 1335

Windfoil,
Rash's Green Industrial Estate,
East Dereham,
Norfolk NR19 1JG
tel: 0362-67115

York Truck Equipment Ltd,
St Mark's Road,
Corby,
Northamptonshire NN18 8AH
tel: 05366-3561

Winter Preparation

Antiwax Advancements,
Anchor House,
225 Station Road,
Nether Whitacre,
Coleshill,
Birmingham B46 2JG
tel: 0675-63637
('AWA' fuel filter heater)

Batoyle Lubricants Ltd,
Colne Vale Road,
Milnes Bridge,
Huddersfield,
tel: 0484-653015
('Helios' fuel additive)

Century Oils Ltd,
PO Box 2,
Century Street,
Hanley,
Stoke-on-Trent ST1 5HU
tel: 0782-29521
(fuel additive)

Dalton and Co Ltd,
Silkolene Oil Refinery,
Belper,
Derbyshire.
tel: 077-382 4151
('Le-Freeze' fuel additive)

Chalbar Ltd,
247/249 Watling Street,
Radlett,
Hertfordshire WD7 7AL
tel: 09276-3233
('Econosol' fuel additive)

Exhaust Ejector Co Ltd,
11 Wade House Road,
Shelf,
Halifax,
West Yorkshire HX3 7PE
tel: 0274-679524
(ECCO 'Hotline' fuel filter heater)

Fleetguard,
Cavalry Hill Industrial Park,
Weddon,
Northampton NN7 4TD
tel: 0327-41313
('Thermo Blend' fuel heater)

Freedom Petroleum Ltd,
Broadheath Oil Works,
Altrincham,
Cheshire.
tel: 061-928 2637
('Freeflow' fuel additive)

Highspeed Lubricants Ltd,
27 Front Street,
Acomb,
York YO2 3BW
tel: 0904-793269
('Vivusol' fuel additive)

Holt Lloyd Ltd,
Lloyds House,
Alderley Road,
Wilmslow,
Cheshire SK9 1QT
tel: 0625-526838

Lubetec (West Yorkshire) Ltd,
18 High Street,
Idle,
Bradford,
West Yorkshire BD10 8NN
tel: 0274-617882
('Fuel Guard' fuel additive)

Lubrication Equipment Ltd,
Lees Road,
Kirkby Industrial Estate,
Liverpool.
tel: 051-548 1183
('Scavenger' fuel heater)

Lucas CAV Ltd,
PO Box 36,
Warple Way,
London W3 7SS
tel: 01-743 3111
(CAV Fuel Heater)

Morris and Co (Shrewsbury) Ltd,
Castle Foregate,
Shrewsbury SY1 2EL
tel: 0743-52431
('Scyllan' fuel additive)

Raychem Ltd,
Faraday Road,
Dorcan,
Swindon,
Wiltshire SN3 5HH
tel: 0793-28171
('Thermoline' fuel heater)

Universal-Matthey Products Ltd,
Underbridge Way,
Brimsdown,
Enfield,
Middlesex EN3 7PN
tel: 01-804 8232
('Unipor' fuel additive)

Wynn Oil (UK) Ltd,
10 Eaton Place,
Reading RG1 7LP
tel: 0734-599948
('Ice Proof' fuel additive)

Workshop Equipment

Colindale Equipment,
Unit 7,
The Hyde Industrial Estate,
Colindale,
London NW9
tel: 01-200 7373

Epco Ltd,
Knowsthorpe Gate,
Cross Green Industrial Estate,
Leeds,
West Yorkshire LS9 0SJ
tel: 0532-495731

HTC (Nottingham) Ltd,
9 Regent Street,
Nottingham NG1 5BS
tel: 0602-415181

Kismet Dynaflex Ltd,
Fenlake Road,
Bedford MK42 0EX
tel: 0234-55211

Leslie Hartridge Ltd,
Tingewick Road,
Buckingham,
Buckinghamshire MK18 1EF
tel: 02802-3661

Lions Workshop Equipment Ltd,
21 Denbigh Hall,
Bletchley,
Milton Keynes,
Buckinghamshire MK3 7QT
tel: 0908-71548

Repco Automotive Equipment (UK) Ltd,
24A Wadsworth Road,
Greenford,
Middlesex UB6 7HD
tel: 01-998 1546

Souriau (UK) Ltd,
Knaves Beech Industrial Estate,
Loudwater,
High Wycombe,
Buckinghamshire.
tel: 06285-24981

Tecalemit Garage Equipment Co Ltd,
Belliver Industrial Estate,
Roborough,
Plymouth,
Devon PL6 7BW
tel: 0752-701212

TI Churchill,
PO Box 3,
London Road,
Daventry,
Northamptonshire NN11 4NF
tel: 03272-4461

TI Crypton Ltd,
Bristol Road,
Bridgwater,
Somerset TA6 4BX
tel: 0278-424300

Walter Somers (Materials Handling) Ltd,
15 Forge Trading Estate,
Mucklow Hill,
Halesowen,
West Midlands B62 8TP
tel: 021-501 1077

Index

ADR *see* road tankers
AEC, 147
Ailsa, 148, 149
alternative fuels, 129-31
   ethanol, 131
   methanol, 131
alternative materials, 27-9
anti-waxing additives, 47
   kerosene, 49-50
      bonded premises, 49
      duty payable on, 49
   mixing, 47
   petrol, 50
Armitage Committee
   proposals, 99
   Report, 65
articulated tractive units, 91-108
   conversions
      ERF chassis, 107
      Lyka, 107
   maximum length, 91
artificial Sun *see* climatic wind tunnel testing
Astra van, 80, *Figure 5.5,* 82, 87
Atomic Research Establishment, 129
automatic transmission, 178
   gear changing systems, 19-24
   microprocessor control, 20-2
   pre-selection of gears, 20-2
   safeguards, 22-4
      manual overide, 24
axles
   combinations
      twin steer, 104
   spread, 107
   weighing systems, 197
Azdel, 35

batteries
   low maintenance, 61
Bedford, 78-9
   CF, 78-9
Berkhof, 155
Birmingham Motor Show, 34
BL, 121, *see also* Leyland
Bova, 155

brakes
   air, 61
   disc
      development, 29-31
      linings, non-asbestos, 31
   S-cam, 61
braking equipment, 197-8
Bridport pebbles, 125
British Technical Council of the Motor and Petroleum Industries, 42
BS 2869, 38
BS 3425: 1966, 66, 67, 69
bump rig, *Figure 8.3,* 124
buses *see* coaches

cab design, 35
cetane number, 131
CFPP *see* cold filter plugging point test
chassis lubrication, 198-9
Citybus, 149
climatic wind tunnels
   Orbassano, 134
   temperature variation, 134
cloud point, 37
club of four, 110-12, 113-14
   cooperation agreement, 112
CM *see* Commercial Motor
coaches
   articulated, 152
   Bedford, 149
   double-deckers, 149, 152, 153
   Ford, 149
   market for, 147, 149
   neo-double-decker, 152
   numbers of in UK, 147
   sales of, 147
coke-up, 131
cold chambers, 134
cold filter plugging point test, 37-8
Commercial Motor magazine
   operational trial route, 137, *Figure 9.1,* 138
   Scottish route, 139
commercial vehicle design, 13
computer data logger, 122
computerised testing, 132-3

227

Concept Cargo, 25-35
Construction and Use Regulations
  Amendment No 7, 91, 106, 107
  articulated combinations
    critical dimensions of, 91
    maximum length, 93
    maximum weight, 92
    tandem trailers, 92
    tri-axle trailers, 92
  1968, Regulation, 91-2
  1978, Regulation II, 98
continuously variable transmission, 13-24
  belt type, 13
  components of, 13
  hydrostatic type, 13
  and Leyland, 13-19
  and Scania, 19-24
  tractive type, 13
Crane Fruehauf, 96, 139
cranes, 199
crash simulation, 129-31
C & U see Construction and Use Regulations
Cummins, 104, 139, 145
  L10, 105
  NTE, 104, 320
CVT, see continuously variable transmission

DAF, 100, *Figure 6.6*, 101, 104, 110, 111, 154
  Variomatic, 112
Daimler-Benz, 121, 145
Dangerous Substances (Conveyance by Road in Tankers and Tank Containers) Regulations, 1981
  Regulation 6, 59-60
decibel, 65-6
demonstrators, 137
demounts, 199
Dennis, 149, 157, 158
Detroit Diesel, 139
diesel waxing, 37-50
  prevention, 37-50
directory of manufacturers and trade associations, 197-226
DON 7250, 31
DoT, 179
  inspectors, 181
double drive, 105
Down Licensing, 179-84
  definition, 181
drive lines, 200-1
driving techniques, 167-78
Duple, 158
dynamometer, 121, 134

Eaton Corporation, 172
  twin countershaft transmission, 173
  Roadranger, 169, 177
Ecosplit gearbox 105
elastohydrodynamic oil film, 13
electric vehicles, 87, 88-9
electronic fuel injection, 132
Empressa Nacional de Autocamones, 114
ENASA, 114, 116-17
encapsulation, engine, 73-5
  Leyland, 74-5
  problems of, 73-5
energy storing, 18-19
epicyclic gearing, 14
ERF, 100, 104
  6 x 4 tractive unit, 104
ETD see Société Européène de Travaux et de Developpement
European Economic Community
  noise legislation
    development of, 71
    Directive, 70/157/EEC, 69
    Directive 77/212/EEC, 70
    Directive 81/334/EEC, 70
    proposals COM (83) 392, 72

Fiat, 112, 134
  Ducato, 78
  Fiorino van, 78, 83
  research establishments
    Arbon, 134
    Bolzano, 134
    Orbassano, 134
    La Mandria, 134
    Nardo, 134
    Turin, 134
fifth-wheels, 102, 210
Finance Act, 1982, 181
fire extinguishers, 63
Fleetline, 148
flowmeter, 139
Foden Trucks, 68, 104, 105
Ford, 149-50
  Cargo, 150
  Transit, 77
freightliner, 114
fuel
  additives, 38
  grade, 38, 39
  tanks, 39-40
fuel delivery lines, 43
fuel heaters
  AWA, 46-7
  CAV Fuel Heater, 46
  Thermoline, 47
fuel systems design, 44

Index

fuel tanks
  and debris, 40
  lagging, 42
fuel monitoring systems, 34, 201-2
fuel starvation, 37

gearboxes
  automatic, 13-24
  constant mesh, 172
  float shifting, 172
  gear ratio spread, 167, 169
  mechanical, 24
  range change, 171
  Roadranger, 169, 177
  servo-assisted, 174
  skip shifting, 177
  splitter, 169, 171, 173
  syncromesh, 172, 175
  torque capacity, 167, *Figure 12.1*, 168

hazardous loads, 51-64
  Code of Practice, 52-60
    chassis construction
      air intake system, 55
      cabs, 53
      electrical system, 56
      engine, 53
      exhaust system, 55
      fuel system, 53, 55
      rear-end protection, 56
      tanks, 58-60
      vehicle stability, 61
Health and Safety Executive
  Guidance Note GS 26, 62
Hendrickson bogie, 104
Hestair Dennis *see* Dennis
hot chamber, 134
hot fuel, 139
Howell, David, 179

ignition improvers, 131
impact testing *see* Millbrook proving
  ground
Industrial Vehicles Corporation, 112, *see
  also* IVECO
infrared lamps, 134
International Harvester, 114
IVECO, 100, 112, 121, 134

Jonckhere, 155

kingpins, 96, 101
Klöckner-Humboldt-Deutz, 112

Lancia Veicoli Speciali, 112, 134
laser doppler anemometer, 129,
  Figure 8.7, 131

lateral braking, 143
layshaft gearbox, 177, *see also*
  gearboxes
leaf springs, 104
Leyland, 68, 147
  automated engine test centre, 132-3
  testing capacity, 132
  Landtrain, 118
  National, 148
Leyland Technical Centre, 121-4
  brake rig, 123
  chassis dynamometer, 122
  environmental chamber, 121
  noise measurement area, 124
  ride simulator, 122-3
  semi-anechoic chamber, 121-2
  test track, 123
light van selection, 77-89
liquefied petroleum gas, 87-8
  advantages of, 88
  availability of, 88
  disadvantages of, 88
London Transport, 148
LPG *see* liquefied petroleum gas

Mack Trucks, 113-14
  Midliner, 113
  MIDR.06.02.12, 14
  MS 250, 113-14
Magirus Deutz, 110, 111, 112
MAN-Volkswagen agreement, 116
MAN-VW, 107, 156
Marston Radiators, 69
Mathweb, 34
MCW, 154-5
  Metroliner, 152-3
  Metropolitan, 148
  Mk II Metrobus, 148
Mercedes-Benz, 100, 156
  1625 S, 105
  2025 S, 105
  2028 S, 105
Millbrook proving ground, 124-9
  acreage, 124-5
  cross-country course, 127
  impact testing, 128-9
  pavé section, 127-8
  steering pad, 125
  test hills, 127
  water trough, 126
MIRA, 138, 142, *Figure 9.4*, 143, 144
Molochite filler, 35
Motor Industry Research Association, 68,
  see also MIRA
Motor Panels Ltd, 34-5

National Bus Company, 73, 148, 152

National Engineering Laboratory, 69
National Express, 152
Neoplan, 153, 157
New Bus Grant, 147
noise, vehicle
　legislation on, 65, 66-89
　sources of,
　　engine, 72
　　turbocharger, 72
　testing procedure for, 66-8

oas, 107
oil companies, 202-3
Olympian, 147, 148, 152
on-board weighing devices, 33
　Loadax system, 33-4
　Concept Cargo, 33-4
operational trial *see* road testing
overdrive, 169, *Figure 12.2*, 170

Paramount, 158
Park Royal bodyworks, 147
Passenger Transport Executive, 148
Perkins, 131-2
Plaxtons, 153, 158
power curve test programme, 133
power steering, 31-2
　proportional feel system, 32
　TRW/CAM Gears system, 32
power take-off, 203
power to weight ratio, 60
Primrose Trailers, 106
prototype vehicles
　testing, 133
psv exports, 147, *see also* coaches
psv manufacturers and distributors, 203-5

Quest, 150, 150-1
　VM, 151
Quiet Heavy Vehicle Project, 68-9
　cooling system, 69
　Rolls-Royce Eagle engine, 68-9
　silencer design, 69
　target noise level, 68

rear-end protection, 34, 63
refrigerated vehicles, 205-6
regenerative braking, 18-19
Renault, 113, 121
　Master, 79, 83-4
　Trafic, 79, 83, 85
research and development, 121-35
restart gradeability, 144
Ridley, Nicholas, 179
road altimeter, 146
road speed limiters, 206-7
road tankers

access ladders on, 62
ADR, 51
brakes, 61
chassis, 60
electrical system, 61
engines, 60
　Regulations, 51
safety, 61-4
standardisation, 61
Roadtrain, 34, 102, 104
road trials *see* road testing
road testing, 137-46
　acceleration, 142
　articulated combinations, 143
　brakes, 143
　computerisation of, 145
　fuel consumption, 139-41
　　measurement of, 141
　speed, 143
　turning circle measurement, 144
rolling road test, 143, *see also* MIRA
Rolls-Royce, 68
Royal Tiger, 154, 158

safety devices, 207
Saviem, 110, 112, 113
Scania, 100, 117-18, 145, 157
　computer-aided gears, 24
Seddon Atkinson, 107
　takeover, 114
'seed vehicles', 137
self steering axle, 101-2
Sherpa van, 81, 83
Sleeper cab conversions, 207-8
Smit, 155
Société Européène et de
　Developpement, 112
STRASS, 145
Strathclyde PTE, 149
suspension systems, 208
　air, 105
　Airpoise, 97
　Granning, 98-9
　leaf spring, 104
　torsion bar, 104
　Volvo, 101
synchromesh *see* gearboxes

tachograph, 141, 146
tail lifts, 208-9
Talbot Express, 78
tank materials, 60
Tenform, 35
38 tonne conversion specialists, 209
tipping gear, 209-10
Titan, 148
trade and professional associations,
　210-12

Index

trailers *see* vehicle conversion
trailer manufacturers/bodybuilders, 212-19
transmissions *see* gearboxes
transmission developments, 13-24
Department of Transport, 91, 98
TRASCO, 145
Trathens, 152
tri-axle trailers, 106, 107, 108
truck building, 109
Truck of the Year Award, 1983, 113
turbochargers, 73, 87
   installation, 73
   maintenance, 73
   manufacturers, 219
   noise, 73
      remedies for, 73-4
   service centres, 219
Type Approval, 180
tyre manufacturers, 219-20

VAG
vans
   diesel engined, 86-7
   four wheel drive, 89
   front wheel drive, 81-2
   rear wheel drive, 82-3, 83-6
van design, 78
   comfort, 80-1
   cubic capacity, 79-80
   height, 79, 85
   payload, 80
   reliability, 81
van Doorne, 114
variator unit design, 13
VEDAC, 34
vehicle control systems, *see* VEDAC
vehicle conversion, 92-101
   tandem trailers, 92, 96
   tri-axle trailers, 93
vehicle finance, 270
vehicle manufacturers, 220-23

VMS, 145
vehicle test tracks, 121-29
Volkswagen
   Golf van, 78
   LT van, *Figure 5.6*, 83
   Transporter, 84-6
Volvo, 110, 112, 113, 149
   Gothenburg model, 100
   MVA department, 101
   6 x 2 tractive unit, 101

washing equipment manufacturers, 223-4
weights and dimensions appendix, 185
wheel distortion, 161
wheel nuts, *see* wheel studs
wheel security, 159-65
   Institute of Road Transport Engineer's Study, 161
wheel studs
   breakage, 159
   clamping forces, 160
   coned fixings, 159
   cracks in, 159
   spigotted fixings, 159
   yield point strain, 160
White Motor Corporation, 114
wind deflectors, 26
   manufacturers of, 224
Windfoil, 26
Wilson Committee Report, 66
winter preparation *see* diesel waxing
   product manufacturers, 224-6
workshop equipment manufacturers, 226
World truck concept, 109-19
   advantages of, 109

YNT, 151
York Airpoise, *Figure 6.4,* 97
York Trailers, 106, 139

ZF gearbox, 117
Zintec, 34-5

Index of Advertisers

Alcan Transport Products Ltd, 98
JWE Banks & Sons Limited, *facing page* 33
Bedford Commercial Vehicles, 2
John R Billows Group, 125
Burmah Castrol Industrial Ltd, *facing Contents*
Eaton Limited, 12
George Fischer Sales Limited, 164
Ford Motor Co Ltd, *between pages* 32-3
Gleason Power Systems, 23
Gulf Oil (Great Britain) Ltd, 43
Hartshorne Motor Services Ltd, 53
Hefac Engineering Limited, 39
Iveco (UK) Limited, back cover
Kysor Industrial (Great Britain) Ltd, 206
Leyland Trucks, 57
Lucas CAV Limited, 45
Lucas Kienzle Instruments Limited, 6
Mercedes-Benz (United Kingdom) Limited, 115
Norde Suspensions Ltd, 5, 99
Octagon Recovery Limited, 41
Redment Engineering Ltd, 90
Rolls-Royce Motors Limited, *bookmark*
Rubery Owen-Rockwell Ltd, 16-17
Ryders International, 59
SAB Automotive Co Limited, 128
SHORFAST Ltd, 19
Scania (Great Britain) Limited, 95
Spicer Drivetrain Group, 21
Triotronics Limited, 55
Volvo Hydraulics Limited, 8, spine
Jonas Woodhead Limited, *facing page* 32
York Trailer Company Limited, 103